Re-Architecting Application for Cloud

An Architect's reference guide

Ashutosh Shashi

Re-Architecting application for cloud
Copyright © 2020 by Ashutosh Shashi

ISBN 978-1-7352222-0-2

for Neha, my wonderful wife.

Table of Contents

About the Author

Ashutosh Shashi is a visionary and innovative technologist, currently living in Atlanta, Georgia, USA. He has completed master's degree in Computer Applications. He has more than 15 years of experience in IT industry.

He is TOGAF 9.1 Certified, Google Cloud Certified Professional Cloud Architect, Microsoft Certified AZURE Solutions Architect Expert, AWS Certified Solutions Architect – Associate, Project Management Professional (PMP), DataStax Certified Professional on Apache Cassandra, and Oracle Certified Java SE 8 programmer.

One can define him as a deep reader, dedicated writer and technology passionate person. He loves to keep himself updated with the latest stack of technologies and industry standards.

Introduction

This is a reference book for Architects. This book can be helpful for those developers who wants to increase breadth of knowledge about tools and technology. If you are planning for career advancement and you are interviewing for cloud architect, this book can also be used for interview preparation purpose. You can go through this book before your interview every time, so that you will remember all the concepts before interview.

As the technology is evolving very fast, new tools and technologies are coming every day. This book covers fundamental of architecting or re-architecting of the application. This book also makes you aware and provides details about tools and technology available in cloud. This book does not over explain any concepts, keeping in mind that you can complete your reading in less time. With this book, you will get lot of information in less reading time.

1

Cloudification

There could be n number of reasons behind the decision to move any application or product to the cloud. Below are the most common reason or factor to rearchitect and migrate the application to the cloud.

a. Business agility
b. Global reach
c. Data analytics
d. Scale to meet on demand

e. Cost
f. Technology
g. Safety

Business Agility:

Business agility provides organization the capability to response to the changes in business environment rapidly and cost effectively. Business agility derive IT agility in any organization, and IT agility derive business agility vice-versa.

Architect should work with businesspeople and users to smartly identify the products which requires change to derive business agility. If you have a product that is dealing with critical market research, how that product will respond with changing market condition.

Sometimes it is absolute necessary to re-architect the application and move application in cloud environment. You decided to re-architect application because the technology used in current application is absolute and any modification in application is becoming very difficult. You decided to re-architect because the third party software that is being used in application is no longer available or no longer supports and you got trapped with that third-party software that is not allowing to enhance your application and you need to end that dependency to develop the capability yourself. You decided to re-architect because your application is not sustaining the increased load because of increase in business and you need to have autoscaling in place, now you are planning to get the feature of cloud computing to scale out and scale in your application automatically. You decided to re-architect because your application needs to support mobile and handheld device that it is not supporting right now, and the application should be available all the time.

To improve agility while moving to the cloud you must re-architect the existing applications, lift and shift or near to lift and shift with little modification in application and application configuration usually do not provide business agility and you will be only able to migrate your application to the cloud with no other advantage.

Global Reach:

If your business is increasing its footprints from local customers to global customer, your application should be accessible from customer base in low latency and you need to have your application and data center in more than one region near to your customer base. To comply with the regulations, you may need to separate the customer data based on the boundary of state or country and store the customers data into their state or country.

To achieve this, you need to re-architect your application and migrate your application to the cloud.

Data Analytics:

Data analytics and market research is becoming the key to the business growth. Every organization is doing data analytics and market research to explore the market condition, to explore the customer need and customer behavior. If your do not have the existing system for data analytics and market research, or if you have system for data analytics that is not fulfilling the purpose in the more competitive market condition and you want to modernize your current data analytics application. You need to understand the data analytics requirements and re-architect your application and move to cloud to get existing data analytics capability from cloud provider and save your time for development and business plan.

Scale to meet on demand:

Scalability of applications needed to perform well in increasing workload situation. Your application is getting popular day-by-day, and the application that is deployed currently on your data center does not support auto scaling. Your operation team is calculating load almost daily and increasing the server capacity or increasing the servers with the load balancer. You do not want to keep calculating the server load on daily basis and keep increasing the server manually.

You need to re-architect your application and move to cloud so that you will get auto-scaling capability.

Cost:

Cost is major factor which enforce organization to re-architecting their application for the cloud. Moving your application to the cloud may not be cost saver for you, you may end up paying more than what you are paying right now. But if you calculate business gain, consider reduced operational expenses, and market reach, then you will realize the benefit of re-architecting your application and moving to the cloud.

Technology:

Technology advancement is a very impartment factor to re-architecting application for the cloud. If you have legacy application with old technology, it is very hard to find good resource, also it is very difficult to do the enhancement or doing any bug fixing in that application. Re-architecting the application using latest technologies and moving to the cloud will provide you flexibility to try out latest technologies to modernize your application over the time.

Safety:

Re-architecting and moving your application to the cloud will allow you to keep your application updated to follow latest security standards. You will be able to keep updating your application with changes in latest governance guideline for application security.

Re-Architecture Decision:

You have a legacy application having many components and using many legacy technologies. Based on organization initiative and business needs you have selected this application for re-architecting for the cloud.

To re-architect, first you have to understand the business and business needs that is being solved by existing application. How many other applications is dependent on this application and how to solve those dependency so that dependent applications should not be impacted. Sometimes you need to understand how those applications are impacting your application and what data and services those applications are sharing with your application and for what purpose. Sometimes you don't know if any application is dependent on your application. For example, your application is storing data in multiple databases, multiple tables and datasets, and some other application is generating report out of stored data that is being stored by your application, or some application may be considering that data for some analytics and research. It is required to understand what is the use of data that is being produced by your application. Create a spreadsheet and list down all the dependent application and how it is dependent on the current application so that when you re-architect your application you should understand the impact of changing service, changing any contract, or changing any database schema or changing the complete database. You may need to refactor dependent application in order to re-architecting your application or find workaround.

You can also list down all the dependent third-party application/software. Check if the same third-party services are still in the market. What business problem these third-party tool and services are solving and still relevant? You should know what processing is being done by third party tool, or services. Make sure you are not trapped into vendor-locking situation before re-architecting your application. If you are in kind-of vendor-locking situation, you should have complete picture of these third-party tools and services, and you should know how to replace it or how it will support the application even after re-architected and after moved to cloud.

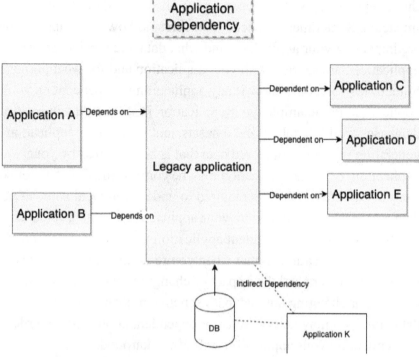

Figure: showing dependency around application.

Now analyze the application, understand the pain points, understand the vision behind this application moderation. Also, it is important to understand architecture practices and framework is being followed in your organization. Go through all organizational architecture guidelines and best practices. List down how these guidelines, and best practices will affect your re-architecture process.

Once you have clear understanding of future application, and expectation from future application, you can do the gap analysis. Investigate about current state of application and its association, its business value, learn about future state of application, understand the gap and do the gap analysis. This gap analysis report will help you when you will start re-designing your application or re-architecting your application.

Next is requirement understanding. Understanding business needs, understanding scaling requirements, understand how it will help organization to grow faster. Calculate return on investment. I have seen architects over-architecting the application, that sometime do not complete in budget and do not provide any extra benefits after investing lots of extra money.

Requirement understanding and stick to that requirement is important while re-architecting the application. Here gap analysis that you have done will help you to understand what is the gap that you need to cover, so that you will not go into details that is not needed.

Now next is deployment model, what kind of deployment model best suited to your application and align with the organization vision and best practice. Application can be divided into small services based on the sub-domain. Each application (group of services) can be designed based on the sub-domain identified, and each application can have separate deployment model. Some can be deployed on bare metal and some can be serverless, combination of IaaS, PaaS, Managed Kubernetes, SaaS, and serverless could be possible.

It is important to understand shared responsibility of cloud services to understand what to consider while designing application. You should understand your responsibility and cloud vendor's responsibility while designing for security.

2

Digital Transformation

Organizations are changing the way of working, changing the way architects used to design applications, changing the way how operation team infrastructure team and developers works. The biggest challenge for organization is to train their people to adopt the new defined best practices, that will evolve gradually as an organization culture.

Best Practices:

Best practices are very important and should be properly defined and followed by any organization. The design process and design approval process should also be clearly defined. Your organization should have architecture governance team. You should know which path to choose, which technology to choose, and what is the organization vision that you are trying to achieve with re-architecting the existing application. You should know and you should have checklist to select the application that needs to be re-architected to achieve organization vision.

The one of the most important tasks for organization is to train the people and help them learn new technology. You can have separate training initiative to train people about new technology, organization process, and about the defined best practices. You need make training a continuous process as technology is changing very fast and all the stockholders, including developer, architects, and managers should be trained properly. Make learning a culture and include it in best practice. Learn from own accomplishments and own failures, write success story and highlight your failures as well. Analyze what you could do better and put in your continuous improvement plan.

There should be common organizational best practices and architecture policy, security policy, a common defined step to follow while making architectural decisions. If possible, create samples and POCs and put them in central repository. Create documents to explain your architectural decisions and log every architectural decision you made with the reason as comment. Document the design options you had and why you considered the current design and why you eliminated the other design options.

Every architecture decision should be taken based on fact. You should have some fact, some data, or some POC to prove your design decision. Architecture decision should not be taken just because of your feeling or opinion about it. Log all the work, POC, or any other research you did that helped you to make some design decision.

Centralized Architecture Governance:

Organizations need to have a centralized architecture governance, a team of architects who is responsible for creating and maintaining guidelines and standards, who will approve all the architecture decisions and any change in architecture.

If any organization is using decentralized authority for architects and there is no co-ordination between them there is a strong possibility of inconsistency of technology stack as each division will think differently and will proceed with different technology stack. This will be an issue for organization as a whole in long term and affect resource allocation, hiring, switching resource between division, organization training program, and lot of other areas as well that will increase direct and indirect cost of product. There will be inconsistency in architecture choices, inconsistency in best practices between various division. Each division will have their own view of application development and design. Maintenance cost will increase over time. If organization will plan to use data analytics that needs to combine different divisions, it may be hard to implement. Separate division will have their own division's data ownership and they design database schema and data model in their own way. They have not coordinated as they didn't know at the time of designing schema that they might need to co-ordinate in future. Security implementation will be inconsistent as each division will handle regulatory complains in their own way.

Considering all the above points, it is very important for any organization to have a central authority who will approve all architecture decisions, best practice implementation, handling of regulatory compliance, and that central authority will have ownership and broad view of organizational concerns. It will increase consistency and reduce cost.

There may be Some challenges with central architecture governance, that is waiting time of approval of architecture, design, and technology stack. To overcome this, there should be guideline and checklist for architecture, design, and technology stack decision. If any division will follow all the guidelines, check all the checklist items, they can send the architecture for approval to central team, who can simply approve that architecture by seeing that it is aligned with the guidelines, all checklist is checked, and followed all the best practices.

Business Strategy:

It is very important to determine business strategy and organization leaders including enterprise architects should have clear understanding on business strategy, goal, and outcome. Before investing into digital transformation, a complete business plan, roadmap, market research, and cost and benefit analysis should be completed. Without completely understanding your goal you should not proceed with digital transformation.

Once you have business strategy in place, you should plan how people will get trained to get align with the business strategy. You can create roadmap for continuous training plan, create artifacts and put it in central repository so that all stakeholder can review it and should have enough information about it.

Architects should understand current business, and also business plan, business strategy to understand future vision of business. Once they have understanding of current and future state of business, they can find the gap in current and future state and do the gap analysis. Also, it is important to analyze risk while doing gap analysis and planning for implementations to fulfil those gaps.

Business risk determination should be a separate process, that will tell us about risk we have, and also analyze how to handle those risk. Architects should understand and create document about the classification of risk and risk mitigation plan.

Data Strategy:

Organization need to prepare data strategy before planning for digital transformation. Data strategy should explain the strategy to manage data and what changes in data model and data flow is needed to achieve business goal.

Data strategy should explain how to manage the data, how to control the data, how to access the data, how to analysis the data. Organization should understand data standards, data compliance, regulatory related to the data, and category of data before defining data strategy. Architects should understand the current state of data architecture and future state of data, he should find the gaps between current and future state of data and do the gap analysis. Strategy about data security and data audit should be clear and defined.

Before digital transformation and moving to the cloud, architects should categorize the data, geographical distribution of data, as well as geographical restriction on data to comply with regulations.

Technology Strategy:

Technology strategy or information technology strategy is pivotal for digital transformation. A good technology strategy should be in place to fulfil the objective defined by business strategy and to achieve business goal. Organizations need to understand their current technology strength, what they want to achieve using technology, what cloud be prospective technology strength, how much effort needed to fill the gap between current technology and future technology.

Public cloud vs private cloud vs multi cloud vs hybrid cloud decision is part of technology strategy. You should determine primary technology stack and define the best practice which will help technology selection. How technology will help business, how much investment on technology is needed, and prediction of business growth because of technology are the part of technology strategy. Technology strategy should be defined at organization level, product level, and project level.

Re-architecting vs Refactoring:

Re-architecting and refactoring of application is two completely different things. If you are moving legacy application to cloud, making it scalable, resilience, breaking into cloud-native microservices, you are re-architecting of application. In re-architecture process of application, you usually re-think about every aspect of application and change the way application works. You change the way application deploy. You change the way of production support. You change the way how it accesses and store the data. You change every side of application area to make it more efficient. You make it more reliable.

You make it faster. You make it more maintainable. And you make it work in more automated way.

In other hand refactoring is changing in code to make application more efficient. In refactoring you just change the code not design. You may change low level design and flow, but you will not change overall design. You change the code, but you cannot change the behavior of application, that is what refactoring is. Never confuse between re-architecting and refactoring of application.

Refactoring is continuous process that is needed because of continuous review process, or improve the performance of application, or to change some piece of code to solve some business problem, or to change code to fix any defects or any security issue, or to handle exceptions in code, or to add more logs for monitoring, etc.

Digital transformation catalog:

Once you have collected all the information for re-architecting application, you should check with the checklist to make sure you are not missing anything.

- Check if you have been communicated the vision clearly and defined in vision statement.
- Check if you have strategy that is already approved by architectural governance of the organization.
- Check if you have already identified all the stakeholders, and who can share what kind of needed information to meet the goal defined in vision statement.
- Check if already have plan to train staffs on new policies and best practices.
- Check if you have created a non-production cloud environment which can be used by technical staff to write POC, implement the concepts they learned to gain confidence on new skills.

- Check if you have agile project management with short term goals and small teams.
- Your entire team will able to handle DevOps work, if they are not capable, you should have plan to train them.
- You should have central code repository in place.
- Check if you have developed the process to evaluate improvements and progress towards goal within specified timeframe.

For digital transformation AWS defined some general design principle to facilitate good design in the cloud, they are as below.

- Stop guessing your capacity needs. In your data center you need to guess your capacity, and peak time should be considered while planning about infrastructure provisioning. In cloud, you should not be guessing the capacity and you should be able to start with as little capacity as you need.
- Test your system on production scale. In your data center it is very costly affair to have the same infrastructure for testing as production. In cloud environment you should have the same infrastructure to simulate the production load, and once your testing will be completed you can decommission your infrastructure. You can build more reliable application with less production defects.
- Automate to make architectural experimentation easier. With automation you can track changes easily, audit the impact, and revert back to the previous parameter easily.
- Allow for evolutionary architectures. In the cloud, the capability to automate and test on-demand lowers the risk of impact of design changes. This allow systems to evolve over time so that business can take advantage of innovations as a standard practice. This will improve your

ability to deliver the changes as per changing business environment.

- Drive architecture using data. In the cloud you can test your experiments easily by doing POC or by creating MVP applications and easily collect the data. You will have flexibility to take decision of architectural choices based on collected data, and you can visualize how your architectural choices affect the behavior of the workload. This will allow you to make fact-based decision on how to improve your workload and design.

- Improve through game days. You can test your architecture and process regularly by performing scheduled game day in production. This will help you to understand where the improvement can be made to develop organizational event to dealing with events.

You should understand and learn about the wide range of service offerings from various cloud providers. Identify which service and options will work well with your workload and understand how to use them to achieve the goal. You can choose multiple cloud and distribute your workload based on the best offering from different cloud. But plan to keep related workload in the same cloud, as it may increase network latency and will be difficult to define networking between cloud. If not properly planned multi-cloud will become more expensive and more problematic.

To select the database, understand the different characteristics of data in your workload. Understand the transactional requirement of data, how you want to interact with data, what is the expected performance in terms of availability and consistency, etc. Decide the type of database based on the research you have done, like relational, NoSQL, etc. Evaluate the best available database service option available based on your database type decision. You can combine different cloud option, like you can use BigQuery from Google Cloud and Amazon RDS from AWS.

3

Microservice Architecture

Microservice is a small independent stateless service that serves one single responsibility based on domain or subdomain. Microservices are responsible to accomplish one small task. It can be scaled independently. Microservices should be able to communicate with other microservice or any other system.

When we talk about re-architecting legacy application for cloud, most of the time we use microservices architecture to re-architect the legacy application.

Reporting Application:

Let's take an example. Company XYZ is having reporting application "ReportIt". ReportIt generates reports in different format and with different data calculation for different purpose. It can read data from customers database and do the calculations for different kinds reports, like sales report, profit margin report, inventory report, payment reports, tax report, customer activity report, etc. Each report can be generated in excel sheet format, graph format, detailed text format, json format, etc. One business division was generating all the reports in different format to analyze and take decision. Now company is growing fast, and it is difficult for one business department to handle everything. Company breaks the business department into separate departments like, sales department, tax department, inventory control department, customer relationship department, revenue department, etc. Data is increasing day by day because of fast growth of the organization. Every department is requesting for multiple reports based on different indicator and in multiple format. Several new data schemas and table get introduced. Several new report formats were also introduced.

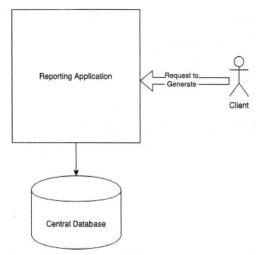

Fig: Reporting application

Reporting application was developed using java, runs over tomcat server, deployed on-premises. Any change in this application goes through change request approval. Builds gets deployed on development servers happens upon request and is not a regular, automated process. Once code changes are done, developer requests for dev build. Once dev build request gets approved, SCM team setup all the parameters and run the build for Dev server. After successful deployment of build on dev server, developer and testing team tests their code and a series of build may be requested for continues defect fix during the testing time. Once developers and testing teams are satisfied with the certain build and there is no known defect in the build, they request for the same build to promote to the staging environment. After staging build deployment completes, testing team starts testing in staging or any other promoted environment. The cycle continues from developer to git repository to dev build to staging build. Once everything looks good, they plan for production deployment.

For production build you usually open a ticket and business team approve it with an agreed production deployment window, at that time you can bring down the application for deployment. Dev, testing, and release team should be available during that production deployment window, so that they can analyze and fix any issue occur during or after deployment.

Now let's discuss about the all the problem organization is facing while using this application. As organization is growing and expected to grow even faster. You should identify all the issues and provide long term solution. Currently almost all divisions are facing issues while generating reports. Sometimes scheduled reports are not working because application crash very frequently. You have vertically scaled the server several times, still application is facing issue and not capable to cater the load because of increasing demands. Here is the list of issues all divisions are facing:

1. Fixing defects and deploying application is making all operation on halt.
2. While fixing defects for one division, application halts for all divisions.
3. If you are fixing defects for one report, it halts application for all kinds of reports.
4. Application is crashing very frequently; and it halts all scheduled reports.

Business is looking for long term fix for these issues. Organization came up with the following plan for this application:

1. Move this application to cloud.
2. Re-architect this application before migrating to the cloud.
3. Re-architect application using microservice architecture.
4. Make this application highly available.
5. Eventual consistency is fine.
6. Remove dependency between divisions.
7. Remove dependency between different reports.

Now you need to think how to re-architect this application. What deployment model is best for your application? Whether to go with serverless, bare metal, or somewhere in between? You can have lots of what and why questions. Once you know all the questions and find out the answer of all the questions, you will design or re-architect your application in better and efficient way.

Breaking the application:
Decomposing application into to small unit which can deploy, and scale independently is called re-architecting the application by following microservices architecture. After decomposing the application, you will decide how the small services will communicate, how the application will react upon any failure, how application will get deployed, how application will get maintained, etc.

To decompose you can think about domain, sub-domain, issues that you are trying to solve, or the problems that you want to solve with this re-architecture. You should know the final goal and expectation from the new re-architected application. You need to extract all the functionality of current application and understand how current application works, what are the feature current application provides and how the current application affecting your business. You should think about database as well, you need to understand the data current application uses. You also need to understand the relationship between different datasets. You should explore how to make data uses and data store more efficient, what will be impact if you change your dataset and decompose them as well. If you will keep one central database system, then it will be the bottleneck for all the microservices. You should think how to distribute the datasets and how microservices will work on their own datasets.

You also can think of events in your application and consider the handling those events in better way.

Below are the possible functions of your reporting application.

Fig: Components of reporting application

You have identified all the components that can be developed as separate microservices and you have explanation of each component, that what it will do and what area of functionality each component will cater.

1. Report Scheduler: this module accepts type and format of report with data and time as an input and schedule the report. You should be able to schedule as many reports as you want. This application will show you scheduled reports as well. If you are scheduling same report for same time or nearby time it will tell you about earlier schedule, and also will ask you if you want to overwrite any previously scheduled report with new schedule. You can cancel your schedule, update your schedule, setup repetitive schedule.

2. Report Data Fetcher: this module is responsible to collect data from various databases and tables that is required to generate report. This module is lots of condition, based and division and report type, and it finds data from various databases and tables based on conditions and insert those collected data into the reporting tables. Reporting tables are the table that stores report data collected by data fetcher. Each data will have report id

associated with it. So anytime any user wants to regenerate the same report data should not be fetched again. If real data get updated, still regenerate report will use the data from reporting table if user asked to do so. If user, ask to regenerate the same report with latest updated data then it should use data fetcher and store the report data in reporting table with different reporting id.

3. Report Creator: Report creator knows what report is being created. It reads reporting data and do all calculation for report and store it in different reporting table. If request comes to regenerate same report with same data as previous report, report creator will provide the same calculation instead of preparing data again. Report creator will generate data for following report:

   a. Sales Report: This is a sub module that handle all kind of sales report. All sales reports calculation and data get created using Report Creator.

   b. Marketing Report: This is a sub module that handle all kind of marketing report. All sales reports calculation and data get created using Report Creator.

   c. Inventory Report: This is a sub module that handle all kind of inventory report. All sales reports calculation and data get created using Report Creator.

   There might be multiple reports like above three reports and all data calculation is being done through report creator module.

4. Graphical Reporting: This module is responsible to generate report using data provided by report creator and create report in graphical format like, pie chart, bar graph, etc.

5. Excel Reporting: This module is responsible to generate report in excel format using data provided by report creator.

6. Text Reporting: This module takes data from report creator module and generate report in text format.

To start designing microservices, if you look at the modules in monolithic application and plan to create service for each module, then you may be going to wrong direction. Do not think about modules of current system, think about functionality and sub-domain. It is easy to decompose application based on modules you figure out, but it is hard to decompose application into true microservices.

If you see the current reporting application, one application doing everything. Major issue is if there is an issue in one function whole application stopped.

Identify Business Capability:

To decompose application into microservices, identification of business capability is important. Sub-domain vs business capability is in question now. I would say both are the same thing. If you think about different business capabilities of application, you can easily think about sub-domain of an application.

What are the business capabilities of reporting application? Think about it. Business capabilities could be scheduling report, creation of report, and representation of report.

Report scheduling capability could do lot of things we already discussed. It can schedule, re-schedule, cancel schedule, update schedule, add more report in schedule, modify number of reports in schedule, modify reporting input data in schedule, etc.

Fig: Report Scheduling

Creation of report capability if the capability of system that can create any report based on request. For example, if request comes for sales report, it fetches data for the sales report and do the calculations to prepare the data.

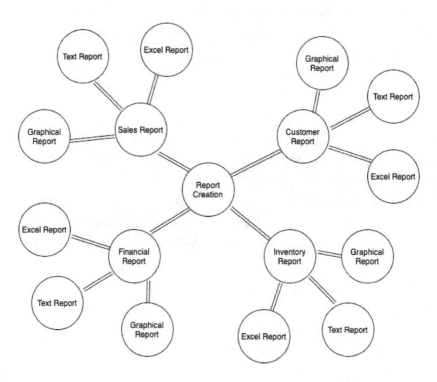

Each report can be represented in multiple format.

Define Microservice:

A microservices is a service which will have all of the following characteristics.

1. Microservice is a small, independent, loosely coupled service that can scale independently. Here scale refers to horizontal scaling.

2. Microservice can be managed by a small development team. A small development team may manage more than one microservice.
3. Each microservice can be modified and deployed independently without updating the whole application.
4. Each microservice is responsible for their own data. Data of each microservice should be separated and not shared.
5. Different microservices can be built using different technologies.
6. Each microservice will expose api, using that it will communicate with other microservice.
7. Internal complexity of one microservice will be hidden from other microservice.
8. Each microservice will have their own separate repository.

Below are the benefits of using microservices:
1. You can do experiments with technology, try multiple technologies without affecting other microservices.
2. If one of the microservice breaks other services will continue serving and complete their task.
3. Microservice can horizontally scale based on resource need for processing.
4. Code maintenance is easy as each microservice will have one functionality and it code size and code complexity will be much smaller than monolithic application.
5. Team co-ordination will be better as small teams will work on each microservices.
6. Easy to fix defects and deploy independently. Provides more agility.

There is some complexity around microservice architecture for architects.
1. You should decide how microservice will communicate with each other Creating each microservice is looks easy but

designing the whole system is way more complex than designing monolithic application.

2. Testing of each microservice would be difficult. Not all the microservice tested independently in real life situation.

3. Network latency is another point architect should consider. Moving big chunk of data between microservice could be big pain.

4. Architecting around data consistency would be complex as each microservices will be responsible for their own data. Data consistency for entire application would be difficult.

Size of Microservice:

Size of microservice cannot be measured by number of lines of code. What does it mean if someone say microservice is not too small and not too big? In my opinion this statement has no meaning. You cannot measure microservice by any logical or physical size or based on any measurement. You can determine microservice by analyzing your application domain and one sub-domain can be implemented in one microservices that has capability to scale independently. Two major criteria you can keep in mind, first is sub-domain and second is scaling together. If you mix both the criteria you will understand you are taking decision of size of microservice by using your business domain knowledge and technical expertise. If your microservice is serving two different responsibility and you scale the microservice horizontally then both the processing does not need scaling all the time. In this case you are scaling microservice because one of the processing need more resource that serves some responsibility, but some other processing some other responsibility is getting scaled unnecessary.

To determine size of microservice first decompose application into microservice based on sub-domain. And then make sure each microservice is serving single responsibility. If any microservice divided based on sub-domain is serving more than one responsibility, then break it again.

I have seen people breaks application based on domain and say it domain driven design. They do not break microservice again even if those microservices are serving multiple responsibilities, thinking they may violate domain driven design principle, and started trapping in bad architecture.

Another consideration is independence of microservices. You cannot make one microservice dependence on other microservices. I have seen people thinking because one sub-domain is dependent on another sub-domain, they make dependent microservice. If you feel the two microservices cannot be independent, try to see if you can make it just one microservice.

Sometimes it is very difficult to have data consistency using two different microservice. You have created two different microservice because you think they belong to two different sub-domains and struggle for data consistency. In this situation merge those two microservices into one microservice.

If your two microservice is very chatty, they talk too much with each other, then check if these two microservice can be merged as it is service one single responsibility and to accomplish that it is doing lots of communication. Combine them together into one microservices in this case.

In reporting application example, once you will decide how many and what microservice will be needed, then measure those service based on microservices principles and make sure you have created microservice boundary correctly and then are loose coupled and can work and scale independently. Then you can decide how they will communicate, it could be service to service communication or event-based system, you can use message queue to make them more loose-coupled.

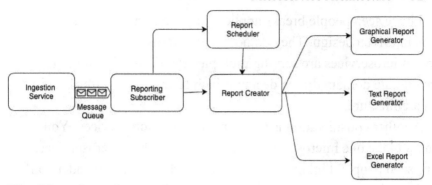

Fig: Reporting microservices.

You know what is microservice and how to decompose the application. When you decide how the microservice will communicate, there are lots of pattern around it.

In reporting application sales division said they want each report in all the format, means all graphical format, text format, and excel report format. They do not want to send the option for format, every time, for every report they want in all the format. Other divisions are having the same requirements.

In above diagram once report creator service will be done creating report data by doing all the calculation, they will call all the three service (graphical report generator, text report generator, and excel report generator) to generate reports in all format.

Orchestration:

Orchestration is a mechanism where a central service will call all other microservice to accomplish the task. In the above example report creator microservice is responsible to calling graphical report generator, text report generator, and excel report generator microservice. If call from report creator to the excel report generator will fail, excel report will not get generated, only graphical and text report will get generated.

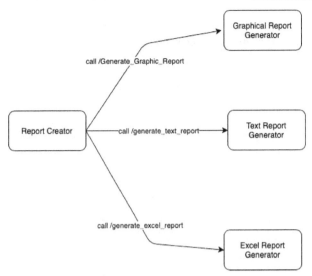

Fig: Orchestration

After call to excel report generator get fail, and you got notified by support team or some log event, you need to do some workaround. You should run the service manually to generate the excel report.

Benefits of orchestration is it is easy to maintain; central service will decide which service to call. It is easy to implement.

There are more disadvantages of orchestration than benefits. It creates dependency, if report creator call to excel service will get failed, it will never be called. Tight coupling between Report creator service and all three report generator services is major drawback. You need to trust on rest call between central service and other services, in this case call between report creator and all three report generator services.

To avoid this dependency, you can use other method called choreography.

Choreography:

Choreography is the way to achieve microservice communication where there is no master service that will call other services. Services will get called automatically. Here each service will work independently and get called based on trigger of certain event. It eliminates the drawbacks of orchestration microservices, services become more efficient less dependent, and more resilient.

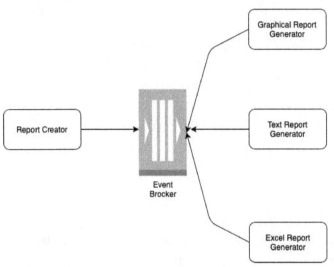

Fig: Choreography

In example of reporting application, report creator service done their job and say I am done. Report creator service is not responsible to call any of the report generator services. Once individual report generator services will see the message from report creator that I am done, each report generator service will start generating reports.

There are several advantages of choreography microservice. You can add more types of report generator service independently without altering anything or without informing report creator service. Similarly, you can modify any report generator service or remove any report generator service. This approach will make microservice faster, and more agile.

Hybrid approach:

Now question is what is better approach? Whether to go for choreography or orchestration style of microservice? And the answer is hybrid. You should not put message broker everywhere. Also, you cannot make dependent service call everywhere. While re-architecting application, keep in mind about benefits and drawback of both the approaches and based on business use case decide which approach is more suitable for microservice communication. If you will see holistic view of your application, you will realize that you need to use both approach while re-architecting application using microservice architecture pattern.

Security in mind:

While re-architecting application for cloud, you should consider security at every step of your architecture. You need to understand the data flowing through your application and what data is most sensitive. Which data you can keep and which data you cannot keep in your application database.

Following is the example of some of the security consideration:
1. Identify all the sensitive data.
2. Classify all sensitive data, highly sensitive to less sensitive.
3. Plan ahead to handle DDoS attack if you are exposing any endpoint.
4. Secure the endpoints.
5. Use API gateway instead of direct call of microservice from client.
6. Use KMS to keep your keys.
7. Use Vault to store any password.

API Gateway:

Each microservice expose one or more well defined endpoints. When client communicate, it can directly communicate to respective microservice, that is called direct client communication to microservice. To make direct call your microservice endpoints should be exposed as public.

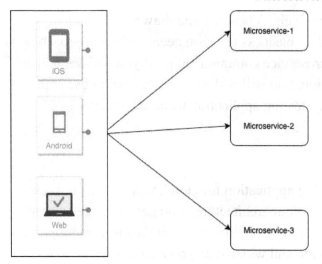

If you think about production deployment, you can have load balancer for more than one replica of microservices, and you will expose endpoint of load balancer publicly. If you deploy your microservice on PaaS or Kubernetes cluster and set your microservice replica to more than one, then it will automatically configure load balancer for you.

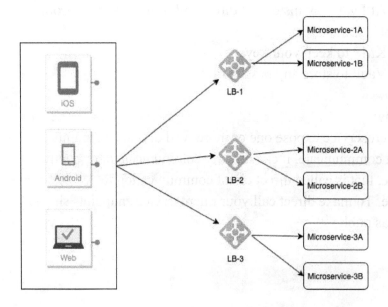

This will work fine for most of the applications. You may face some issue using this model. You have web application from UI to backend, you are directly communicating with multiple microservices, and there are lots of calls between the client and microservices. More and more microservices are communicating with clients directly to open internet, you need to secure more and more end points. You implemented authorization, and security at each microservice level. It will be difficult for you if you need to route your call to microservice dynamically. If you have some microservice that does not support HTTP/HTTPS protocol, it will be difficult to interact with client.

If UI client from open internet is making direct calls to multiple microservice, any change in those microservice may need to update the client side as well and this may happen frequently. The client UI should know which microservice is serving which purpose.

Another approach is to use API gateway. API gateway provides single entry points to the client and manage the calls from group of microservices. API Gateway make it easy for client comminute with microservices as this will be the single point of contact. You can modify microservices, add more microservices, or remove microservices without letting client know. Client will not be aware about any changes in backend.

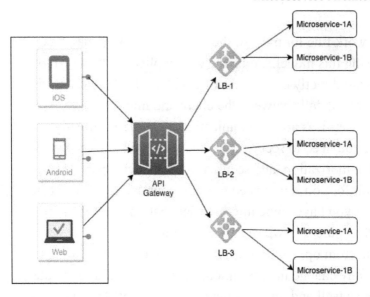

You can update your client and also update any of the microservices independently. You can authenticate clients call at API gateway level. Also, you can implement cache, you can dynamically redirect client calls to the right microservice.

One microservice can be used for mobile app and web client, but if you are doing things differently for different device and different operating system calls, then you may need to have separate API gateways for different platform.

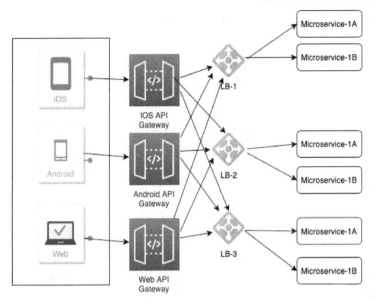

You can have separate API gateway for different platform as shown in above figure. For android app, iOS app, and web you can handle the calls separately.

You need to understand that you cannot have a single API gateway for all of the microservice, you may need to have multiple API gateway based on type of app, type of request, or based on any other criteria suitable for your application.

Benefits of API gateway could be:

1. API gateway can act like load balancer.
2. You can implement retry policies, and setup number of retries.
3. It will help you to detect and block DDoS attack by implementing rate limiting, API quota, throttling.
4. Authentication and authorization can be done at API gateway level.
5. You can implement caching of response.
6. You can allow access to the trusted IP addresses, or trusted IP address range by doing IP whitelisting.
7. You can add headers or add some more request field at API gateway level.

8. Logging and versioning can be done through API gateway.

There are many more benefits of using API gateway, but there are certain limitations as well. API gateway could become a single point of failure. See if you have flexibility to scale your API gateway as well. Also, there could be some delay in response because API gateway is an additional hop.

12-factor App:

12-factor app is very popular application development and deployment methodology, was mainly for SaaS app development and deployment is applied heavily for microservice development and deployment. In 12-factor app methodology, 12 factors are:
1. Codebase
2. Dependencies
3. Config
4. Backing services
5. Build, release, run
6. Processes
7. Port binding
8. Concurrency
9. Disposability
10. Dev/Prod parity
11. Logs
12. Admin process

1. Codebase:

One app (microservice) should have one codebase. Even though multiple version of your application is in production, you must control all the multiple version through single code base repository. You have created service for multiple customer, each customer needs different type of customizations, you can have base branch and then based on different customer, you can have separate branch for each customer. Each customer branch can be main branch for that particular customer, and then you can have multiple subbranch again.

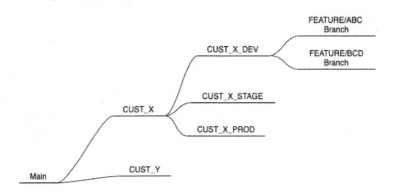

Above is complex scenario where different version of your service is being used by different customer.

In simple scenario, your codebase will have dev, stage, and prod (master) branch, and developer can fork feature branch from dev branch and submit the merge request to merge feature branch to dev branch.

Best practices that you have decided to work with repository should be communicated properly to the developers and have some training documents to follow.

2. Dependencies:

All dependencies on any app (or microservices) should be declared at common place. You should have the list of dependency at one place where you can check and analyze dependency.

For example if you are developing application in node, react application you define all the dependency in package.json.

```
{} package.json  ✕

{} package.json > {} devDependencies
  1   {
  2       "private": true,
  3       "workspaces": [
  4         "packages/*"
  5       ],
  6   >   "scripts": { …
 21       },
 22       "devDependencies": {
 23         "@testing-library/jest-dom": "^4.2.0",
 24         "@testing-library/react": "^9.3.0",
 25         "@testing-library/user-event": "^7.1.2",
 26         "alex": "^8.0.0",
 27         "eslint": "^6.1.0",
 28         "execa": "1.0.0",
 29         "fs-extra": "^7.0.1",
 30         "get-port": "^4.2.0",
 31         "globby": "^9.1.0",
 32         "husky": "^1.3.1",
 33         "jest": "24.9.0",
 34         "lerna": "3.19.0",
 35         "lerna-changelog": "~0.8.2",
 36         "lint-staged": "^8.0.4",
 37         "meow": "^5.0.0",
 38         "multimatch": "^3.0.0",
 39         "prettier": "1.19.1",
 40         "puppeteer": "^2.0.0",
 41         "strip-ansi": "^5.1.0",
 42         "svg-term-cli": "^2.1.1",
 43         "tempy": "^0.2.1",
```

For java bases application, you can have manifest file for dependency management. Maven or griddle can be use for dependency management and having central repository. Your organization should have a central antifactory to have all the dependency stored there.

testproject/pom.xml

```
1⊝ <project xmlns="http://maven.apache.org/P(
2     xsi:schemaLocation="http://maven.apache.
3     <modelVersion>4.0.0</modelVersion>
4     <groupId>org.test.project</groupId>
5     <artifactId>testproject</artifactId>
6     <packaging>war</packaging>
7     <version>0.0.1-SNAPSHOT</version>
8     <name>testproject Maven Webapp</name>
9     <url>http://maven.apache.org</url>
10⊝    <dependencies>
11⊝      <dependency>
12          <groupId>junit</groupId>
13          <artifactId>junit</artifactId>
14          <version>3.8.1</version>
15          <scope>test</scope>
16      </dependency>
17     </dependencies>
18⊝    <build>
19        <finalName>testproject</finalName>
20     </build>
21  </project>
22
```

Above is the example of dependency defined in maven pom file.

If you are creating docker container, you again need to define all the dependency at one place called Docker file. It is easy to modify, add, and delete dependency if it is defined at a central place.

3. Config

Application configuration can have very sensitive information, like database connection string, app credentials, service account credentials, etc. If you store all of this information in configuration file, and checked-in in repository, then it will be accessible to many people who may not be intended to know that sensitive information.

All sensitive configuration should be externalized from source code, to make sure you are not storing it in source repository. You usually have multiple environments where code is deployed, like dev, staging, and production environment. Each environment will have their own environment specific database and database connection string, environment specific service account credentials, etc.

You can store all sensitive configuration information at secure place like vault for each environment. At runtime you detect the deployed environment, like dev, stage, or prod, and fetch the sensitive configuration from secure storage and replace in your code environment with the placeholder. In this way you will not reveal your sensitive environment information. For example, if you are using Kubernetes engine, you can externalize configuration using ConfigMaps or Secrets.

4. Backing services

If your application is consuming any external service, whether it is local or third-party service, it should be attached to the resources. Application do not differentiate between local and external service provided by third party.

Here attaching to the application mean to say, you can add those service url in configuration. Your application will not aware about actual application but the attached service.

For example, your application is using MySql, and Amazon s3 as backing service. You are storing data in MySql and storing files using amazon storage service S3.

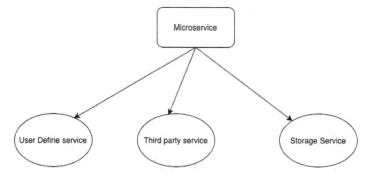

If you want to change Amazon storage service S3 to Google bucket, you just need to update the URL and credential in configuration, and no change needed in code. Application will keep working as is.

5. Build, Release, Run

Build, Release, and Run should be separate process and each of the process should produce uniquely identifiable artifacts.
Once you checked-in code to your source repository, your code should be built in pipeline. If build is failed because of some compilation issue, build issue, or because of any test case failure, it should notify developer with relevant information so that developer will understand and fix that issue in code. Fixing issue and committing the code to source control will cause the trigger the build again. Each time build is successfully built, it should mark that build with a unique build number.
If you plan to release, it should go through some automated process. You can select which build number you are going to release. Each release will have a unique release id. For example, current dev release number is 1.1, the next release number could be 1.2. You are releasing new 1.2 version, and after releasing you found some issue in release that need to fix at code level and then it should have to come for release again. Now your automated process should be able to revert

back the release 1.2 with release 1.1, that was working fine. All the release should always have the option of rollback.

Once you release your application, it gets deployed and should point some environment. For environment, during release you need to attach the environment configuration with the build. After deployment, based on environment configuration you attached with build, it will run for that particular environment. Let's say you have attached stage environment configuration having stage database connection string, stage cloud storage folder credential, etc. Then after deployment this application will run for staging environment.

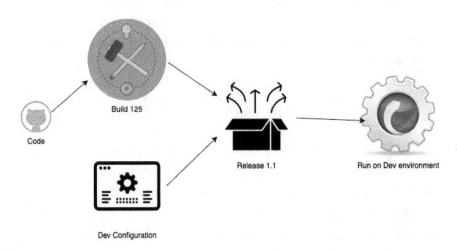

Code

Build 125

Dev Configuration

Release 1.1

Run on Dev environment

6. Processes

One or more services (apps) can be in execution, and all services should be stateless. If you have one microservice running in some execution environment (dev, stage, or prod), you can scale it to multiple processes, but all processes should be stateless. Stateless means, no process can share any kind of data between execution. Services should not keep data in memory to share it between execution.

If your stateless service has to store any data, you can attach database, cloud storage, or similar backing service to your process. In this way you can use stateless apps and enforce loose coupling.

7. Port binding

Deployed applications have to expose services using port binding. Earlier, monolithic applications were deployed in applications containers like IIS, Apache Tomcat, IBM WebSphere application server, etc. Twelve-factor apps does not suggest using external application servers.

Your application should be bundled with webserver library. Your application is completely self-contained and will able to expose HTTP as a service by binding a port.

For example, if you are creating spring boot application for microservice that comes with embedded webserver. You use the best practice to externalize port number to configuration file.

You can abstract port binding by mapping service, if you are using service discovery, or if you are using any load balancer. All depends on how you design your deployment model.

8. Concurrency

While re-architecting your application to migrate your application to cloud, you decompose your application in to small independent microservices. Each microservice should be stateless, and you will be able to run multiple concurrent application to distribute the handle the processing load coming to that process.

Application should never daemonize, or write PID files, instead it should rely on operating system process manager.

9. Disposability

Applications, which is running in cloud environment runs on disposable infrastructure resources. Anytime application can be started and killed for scaling out and scaling in. Application should be able to gracefully shutdown and start quickly to support scaling, and quick release.

When you send SIGTERM signal to the app container, application should start doing all cleanup process, and after completing those cleanup work, it should be stopped. When you scale in your app, some replica will get gracefully stopped.

Your application startup time should be minimal. To reduce startup time, you must check that you are not loading unnecessary dependency for your application. Also, you should manage your environment variables outside of your application, so that your application can consume those variables at runtime and reduce startup time of application.

10. Dev/Prod Parity

Dev/Prod Parity is environment parity between dev, stage, and Production. Your dev, stage, and prod environments should be similar.

If your environments are not similar, you may be fixing some defects in production that could be caught in dev or stage environments. It may take months to move your code to production, and you may lose the time sensitive business opportunity. Developer will write the code and they will have no idea how code is getting deployed. Code deployment is driven by operation team that is completely different team. Also, developer may be using some different tool to test their code and operation team deploy in production with completely different set of tools.

A 12-factor app should be deployed immediately after developer writes the code. Developer who writes the code should closely involve with operation team for deployment and see the application behavior in production. You can have DevOps team and best practices around it.

You can make use of source repository, configuration management, and configuration template to make dev/prod parity. If you are using docker container, you are making sure the same infrastructure across all environment. Same docker image is getting deployed to dev environment and then moving to stage and production.

Source Code

As described in build, release, run, you are keeping builds and release separately. Once your build is successful on dev environment, and you know which build number is successful on dev environment, you move the same build number to release on stage environment. Similarly, you know which build number is tested well and is successful on stage environment, you move the same build number with the same properties, and using the same configuration template and with the same backing services, you will move this same build to production.

11. Logs

Log is a sequence of event application writes in the file with the timestamp and event name. Logs provides you the information about your application behavior, information about any defect, health of your application, etc. By doing log monitoring you can monitor your application.

It is easier to write log file and analyze if just one instance of application is running. When your application scales into multiple instance, you may be generating separate log file for each instance. It will be hard to track all the instance and get log file from each instance for monitoring. In cloud environment your instance may be scale in and scale out very frequently and keeping track of log file is really hard in that situation. You need to have some centralized log management where all of the logs from each instance will get streamed. Streamed log will have information like application name, instance id, timestamp, and other key value pair of log information, so that you can search the log based on application name, timestamp or time range, and other searching criteria.

Almost all cloud provider provides centralized logging services, that you can use to stream your log. Some third-party service providers also provide centralized log management services.

If you are using such tool, you can have all the logs at one central place, and you will have facility to search the log from that centralized log storage. You can also schedule notification if any specific log gets logged, so that you can monitor your application and any specific log may alert you, so that you can go back to your application or respective data to analyze. You can also schedule notification service if any error or exception log gets logged.

12. Admin process

Admin process in 12-factor app is talking about administrator or management task that runs as one-off process. Example is generating reports, run database migration, invoking scripts, creating database backup, etc.

Admin code should be shipped with application code to avoid synchronization issue. Admin-process should also run in the same environment as regular application is running. Admin process should run against the release, using the same codebase and release.

If you are developing and releasing the application, your application can have separate api for admin process with right access permission, or your application can have some startup and shutdown script to achieve some admin or management task.

4

Serverless Solutions

Serverless is a cloud computing execution model where cloud providers provide the execution environment to execute the code. Cloud provider manages all the infrastructure and you only manage the code and execution related stuff.

You can say in serverless, all resources are provided by cloud provider in form of SaaS service model. Operational cost in serverless environment is negligible, you do not have to think about infrastructure, scaling and load. In serverless, there are few limitations, and it follow some rules and suits for specific use cases.

You can have mixture of execution model to achieve any goal, by designing some part of application on serverless and some part on PaaS, IaaS, or Managed Kubernetes, or mixture of all. Serverless is an architecture where most of the operational responsibilities managed by your cloud provider. It allows you to develop application and think about your application, you do not need to think about operating system maintenance, servers, capacity planning and provisioning, creating cluster, or defining scaling rules. You can focus more on your core business, rather than operating and managing infrastructure or runtimes. You can spend more time on your application which can scale based on demand and are reliable.

FaaS (Function as a Service):

Function as a Service (FaaS) is a type of serverless computing where your code is distributed among functions and gets executed in response to events.   FaaS is a specific form of serverless or you can say it is subset of serverless, meaning of serverless is broader than just FaaS.

You write small piece of code called function. Each function will respond to the event. Let's say you have application, that is collection of ten independent function and each function is responding to some event, and your cloud provider will bill you for function execution time only. Each function will get executed on some runtime, on some server assigned by cloud provider. You do not need to know how your code is getting executed, how your code is getting scaled.

Few benefits of FaaS, is the faster way to the market, you just need to write the code, run code without provisioning the server, your process will scale out and in based on load automatically, you will get charged for your code execution time only irrespective of how many parallel process is running (it may differ by cloud providers), you will get consistent performance without planning for it, etc.

AWS Lambda:

AWS Lambda is a FaaS, where you can execute your code in response to an event and AWS will manage underlying compute resources. Lambda provides high availability and perform administrative task related to provisioned environment like, server and operating system maintenance, capacity provisioning and automatic scaling, security patch deployment, code monitoring and logging. The code you run on AWS Lambda is called **Lambda Function**. Once you have created Lambda Function, it is ready to run as soon as it is triggered by some event, such as HTTP requests via Amazon API Gateway, modification to objects in Amazon S3 (Object Storage) buckets, table updates in Amazon Dynamo DB (Managed NoSQL Database), state transition in AWS Step Functions (Serverless Workflow), etc.

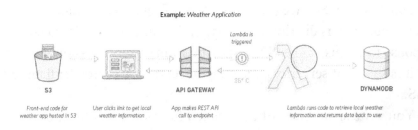

Example: *Weather Application*

Each service that integrate with Lambda send data to the Lambda Function in JSON format as an event, the structure of JSON message depends on event type. Lambda runtime convert the event into an object and pass to the Lambda Function.

The Lambda runtime converts event document into an object and pass it to the function handler. Event source mapping reads from an event source and invokes a Lambda Function. Event Source Mapping can also process items from a stream or queue in services that don't invoke Lambda Function directly. Lambda can read event from Amazon Kinesis data stream, Amazon Dynamo DB, and Amazon Simple Queue Service (Amazon SQS). You can use a resource-based policy to allow AWS service to invoke your function. Depending on the service, the invocation can be synchronous or asynchronous.

In case of synchronous invocation, invoking service will wait for response from the function, and it can retry the invocation in case of error. Here are the services that invokes Lambda Function synchronously:

- Application Load Balancer (ELB).
- Amazon Cognito (User Identity and data synchronization service).
- Amazon Lex (Service for building conversational interface).
- Amazon Alexa (Voice assistance on Amazon Echo).
- Amazon API Gateway.
- Amazon CloudFront (Lambda@Edge).
- Amazon Kinesis Data Firehose (Managed ETL).
- AWS Step Functions (Serverless Workflow).
- Amazon S3 Batch Operations.

In asynchronous invocation, Service which send the event queued and invoking service gets the success response. From queue event passes to the Lambda function, and invoking service is not aware that what happened in processing, whether it failed or successfully processed. In case of failure, Lambda handles retry, logging the error, send failed event to the dead-letter queue based on your configuration. Asynchronous invocation increases agility and allow service to scale flawlessly. Here are the services that can invoke Lambda Function asynchronously:

- Amazon S3 (Object Storage).
- Amazon SNS (Notification Service).
- Amazon SES (Email sending service).

- AWS CloudFormation (Infrastructure as Code).
- Amazon CloudWatch Logs and Events.
- AWS CodeCommit (Fully Managed source control).
- AWS Config (Configuration Management).
- AWS IoT.
- AWS CodePipeline (Managed Continuous Delivery Service).

AWS Lambda has native support for java, go, PowerShell, Node.js, C#, Python, and Ruby. Each Lambda function gets 500MB of non-persistence disk space in its own /tmp directory. All the Lambda functions should be stateless, but through code you can access and store stateful data from Amazon S3 or DynamoDB.

In AWS function, you configure your function by allocating the memory to the function. Based on memory allocation CPU capacity will get allocated. Your cost of function execution will get calculated based on memory you have allocated. You can calculate cost on AWS calculator provided.

AWS Lambda Function can be used to create web app, image or file processing application, data streaming or data processing applications, creating mobile backend, etc.

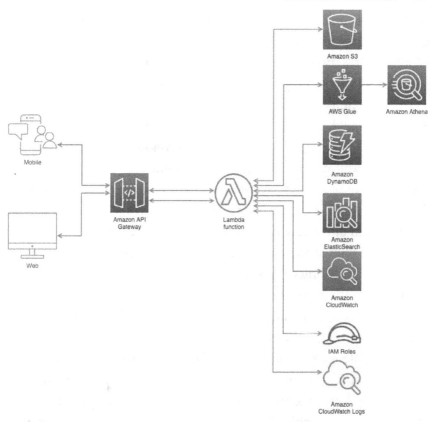

Above diagram is an example of how you can design web or mobile application using Lambda Function combining with other AWS serverless offering.

Google Cloud Functions:

Google Cloud Functions is event driven serverless compute platform similar to AWS Lambda Function. You can use Google Cloud Function connecting with other google cloud services or other third-party services to develop your application.

Google Cloud Function supports only Node.js, Python, and Go. If you are thinking if you could write cloud function in Java or C#, google cloud functions is not an option for you. You can write Google

Cloud Functions in Node.js, Python, or Go, and your function will get executed in respective runtime provided.

You can write two types of cloud functions, HTTP cloud functions, and background functions.

HTTP functions can be invoked via HTTP requests which supports all the request methods like, GET, POST, PUT, DELETE, and OPTIONS. HTTP cloud functions are synchronous, it gets invoked by HTTP request and wait for response from cloud functions. A TLS certificates automatically provisioned for HTTP cloud functions and it get invoked through a secure connection.

Background cloud functions handles event from cloud infrastructure, such as message from a cloud PUB/SUB topic, or changes in Cloud Storage bucket.

Cloud function is having a "name" property that is set at deploy time, and once set you cannot change this property. This property gets used as an identifier for function by google cloud. If your function name is calculate(), you can set "name" property as calculate at deploy time, and calculate will get executed. If you want a separate name other than function name for this property, then you can provide function name –entry-point flag, so that cloud function should know the name function get executed.

Google cloud function supports following google cloud native triggers:

- HTTP Triggers (Trigger through GET, POST, PUT, DELETE, OPTION methods).
- Google Cloud Pub/Sub Triggers (Trigger through message published to Pub/Sub topic).
- Google Cloud Storage Triggers (Trigger through Object creation, deletion, archiving, and metadata update in google cloud storage).
- Google Cloud Firestore Triggers (Trigger through Cloud Firestore create, update, delete, and write event).

- Google Analytics for Firebase Triggers (Trigger through log event).
- Firebase Realtime Database Triggers (Trigger through write, create, update, and delete event).
- Firebase Authentication Triggers (Trigger when user account is created, or user account is deleted).
- Direct Triggers (Usually direct triggers are used for debugging, maintenance and quick iteration).

You can trigger Google Cloud Function from any service which support Cloud Pub/Sub as an event bus. If you want to invoke any other service that does not support Cloud Pub/Sub, then you can use HTTP invocation.

User Cases:

There can be lots of use cases to use google cloud functions, some of the use cases are:

Streaming data from cloud storage (Object Storage) into BugQuery. You can use cloud function for this. You can have cloud functions that will transform the file and send the data to the big query for analytics.

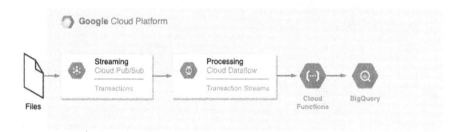

You can use cloud function for device to device communication by integrating it with cloud IoT core module as per below:

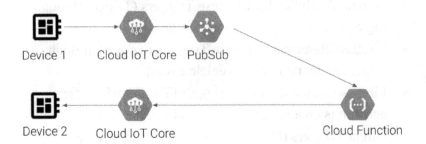

Device 1 will use Cloud IoT Core to send telemetry message to Cloud PubSub. Write Cloud Function that will get trigger when any message gets published on Cloud PubSub and receive the message. Cloud Function do the necessary processing to that message and send message to Device 2 using Cloud IoT Core.

You can create cloud function to serve as:

- Mobile backend.
- Web application.
- Notification service.
- To send email after payment successful.
- Real time file processing.
- Real time stream processing.
- Voice and image analysis.
- Creating virtual assistance.
- Serverless IoT backend.

Google Cloud Run:

Cloud Run is new way to run FaaS kind of service. Cloud Run is fully managed and charge you only for resources you have used, that is somewhat similar to the cloud functions. If you have any application running round the cloud and users are available for that application round the clock then cloud functions or Cloud Run would not be best choice for you, as you will end-up paying more. If your application is kind of batch application running few hours daily, or some office application that is being used in day time only and no one use in night, Cloud Functions, or Cloud Run is good in this case, as you will save the cost for the period application is not running.

Cloud Run is serverless, and fully managed compute platform that automatically manage all the underlying infrastructure. Cloud Run use containers, internally runs on the Kubernetes cluster. Cloud Run is based on Anthos, which let you build, deploy, and manage application either in cloud or on-premises in a secure and consistence manner.

Cloud Run build upon open standard, use Kubernetes based platform to deploy and manage, enabling the portability of your applications.

For Cloud Run, you can write code in following languages:
- Java.
- Python.
- Go.
- C#.
- Node.js.
- Ruby.
- PHP.
- Shell.
- And many others also supported.

You do not have to think about infrastructure, you need to focus on your core business. You will get charge only when your code is running.

Key features of Cloud Run are:
- You can use any programming language of your choice, or any operating system libraries, or even you can bring your own binaries.
- Cloud Run can easily pair with Cloud Build, Container registry, and Docker.
- It is simple to deploy quickly and simple to manage.

- Cloud Run can scale down to zero and scale up from zero to N quickly depending on traffic.
- Cloud Run services are regional, that can easily replicate across multiple zones.
- Cloud Run is very well integrated with Cloud Monitoring, Cloud Logging, and Error Reporting.
- Cloud Run is built on Knative (Kubernetes based platform to deploy and manage workloads).
- Every Cloud Run service will get HTTPS endpoint with TLS termination handled for you.
- If you have custom domain, you can map your cloud Run service to your custom domain.
- Having excellent microservice support.
- You can move workload across, there is no vendor locking.
- It is using GKE, you can enable GKE for cloud run for Anthos.

Use Cases:

Using cloud run you can build website using latest and best technology stack.

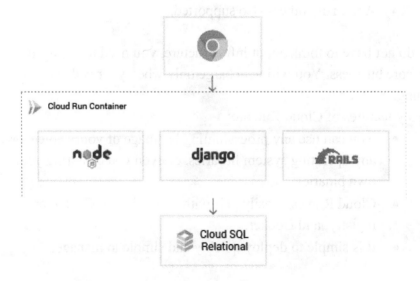

Cloud run can be used to develop mobile backend application that manipulate, store, and retrieve data.

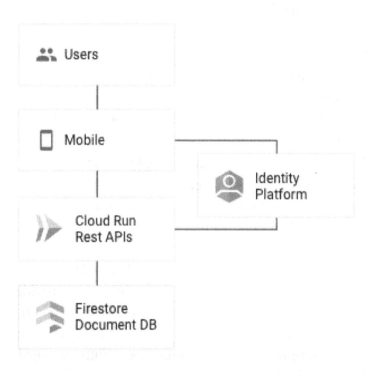

Using Cloud Run you can build containers using data processing libraries and do the data transformation before moving data to BugQuery.

You can use Cloud Run for batch processing which will take data from big query and generate daily report.

Create and Deploy Cloud Run:
- Create an application (e.g.- create a basic web server) that listen on a port defined by the PORT environment variable.
- Create a new file named Dockerfile in the same directory as the source file and define all dependency and runtime in Dockerfile. This will containerize your app in docker image.
- Build the docker image using gcloud build command.
- Deploy your app using gcloud run deploy. Your application will get deployed on cloud run, and cloud run will automatically handle scaling, endpoint, etc.

There are two cloud run platform available, google cloud run (fully managed), and Cloud run for Anthos. Cloud run for Anthos provides access to custom machine types, additional networking support, and cloud accelerators. Cloud run for Anthos will allow you to run your workload either on on-premises or on google cloud. Fully managed cloud run platform will allow you to deploy stateless containers without worrying about the underlying infrastructure, workload will scale up with the traffic and scale down to zero if you are not using it. You will get charge only when your application is running, that is billed to the nearest 100 milliseconds.

Azure Functions:

Azure functions are the Microsoft offering of FaaS. It is similar to the AWS Lambda Functions, and Google Cloud Functions. Azure functions are event driven and compute-on-demand. Azure functions are also following pay when use model, that means you have to pay for the period when your function is running. Azure functions are not meant to for long running processes, if you will deploy long running processes as FaaS, then you will end up by paying more. For long running processes Managed Kubernetes or PaaS is good option. Azure functions version 3.x fully supports following programming languages for production use:

- Java.
- PowerShell.
- Python 3.x.
- TypeScript.
- C#.
- JavaScript (Node.js).
- F#

You can develop Azure functions on Azure portal. Your complex application and team of developers needs to develop the code on local computer, test and debug the code from local computer as well. You can setup local development environment based on your choice of language, once developed, you can connect Azure services from your local code to debug and test your code. You can use pre-defined template to develop your Azure function project. You can use visual studio code to develop Azure function using Node.js, use visual studio to develop Azure functions in C#, or you can use java and maven to create Azure function in eclipse or IntelliJ Idea IDE. Azure functions are stateless and asynchronous.

Durable azure functions:

Durable functions are Azure functions that allows you to write stateful functions. There are two ways you can write durable function and maintain state, one is to define stateful workflow, that is called orchestrator functions, and other is to define stateful entities, that is called entity functions, both the type of durable functions follows Azure functions programming models.

Durable functions are not supporting all the programming languages, that are supported by Azure functions. You can use C#,

JavaScript (Node.js), or F# to write durable functions. Like azure functions, you can use template to write durable functions.

Durable functions can be used, when you need to implement function chaining pattern where sequence of functions get executed in specific order, and output of one function will be input of next function.

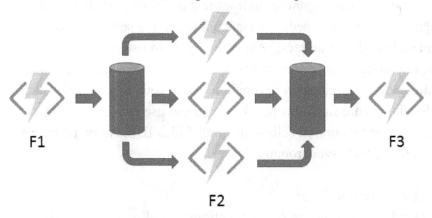

Here F1, F2, F3, and F4 has to be executed in order and input of F2, F3, and F4 is the output of F1, F2, and F3 respectively.

Another use case of durable function is fan out/ fan in pattern where you execute multiple functions in parallel and one function is waiting for all parallel function execution to finish and use the output of all the function executed in parallel as its input.

There are other several use cases as well, like https call from user, etc. You can have stateless functions whenever possible and maintain state in some external storage.

Triggers for Azure functions:

You can invoke Azure function using triggers, similar to AWS Lambda functions, or google cloud functions. Each Azure functions must have exactly one trigger. Trigger will trigger Azure function with some payload, that will be input information for Azure function to execute and process.

Azure functions can be triggered by a queue, when a new message arrives in queue, that become trigger for Azure functions. Azure function process that message as coded and store output in storage or send message to some other queue. You can have Azure function that reads data from some storage and store data in some other storage upon execution, and trigger of that Azure function could a job scheduler or timer. Your Azure functions can be triggered using Azure even grid or using HTTP trigger.

Deployment of Azure Functions:

You can setup continuous deployment of Azure function using GitHub (Azure Repos, and Bitbucket is also supported). GitHub will be your single source of truth where developers write code and it will trigger deployment after any change in branch as per configuration. You can easily setup continues deployment by selecting Deployment Center in Azure Portal and authorize GitHub.

Deployment with Deployment Slots:

You can use Deployment Slots for your Azure functions deployments, so that your traffic redirection will be seamless. Deployment slots purpose is similar to the blue-green deployment. Each deployment slots are running instances with different environment and exposed via publicly available endpoint. One instance will always be mapped with production slot and you can swap instance assigned to that slot on demand. If function is running during swap of instance, then execution continues with the same instance and next trigger will be directed to the new swapped instance, so the request should not drop during swap. After swapping you realized that new production slot is not working as expected and you need to fix something before deploying it again, you can reverse the swap immediately to move to the last known good instance.

Before swapping all slots scale out to the same number of slots as production slots, so that swap will be easy and smooth. The biggest limitation with slots is, it is not available for the Linux consumption plan, but available with the Linux premium, Linux dedicated, and all Windows plan.

Deployment with Azure pipelines:

Azure pipelines can be used to implement continuous integration and continuous delivery for Azure functions. You can use template in Azure pipeline, that is ready made task to build and deploy Azure function. Or you can use YAML file, where you can define build steps and release, and keep the YAML file in the same repo as code.

Deployment with GitHub Action:

You can define workflow with GitHub actions like push. Workflow is an automated process that you define in GitHub repository, that describe how your functions will get build and deploy. You can define workflow in the YAML file in the ". /github/workflows/" directory of your repository. YAML file defines the workflow with various steps and parameters, having three separate section for authentication, build, and deploy.

Deployment of .zip:

You can zip your Azure functions code and deploy directly through CLI or API provided for that purpose. The zip deployment API takes the contents from .zip file and extract into the wwwroot folder of your function app.

KEDA (Kubernetes Event-driven Autoscaling):

KEDA is a single purpose and light weight component that can be added to any Kubernetes cluster and works as horizontal Pod Auto-

Scaler. With KEDA you can specify the app you want to scale in an event-driven way while other apps continue to function.

KEDA pairs seamlessly with the Azure functions runtime and tooling to provide event driven scale in Kubernetes. Your Azure function app should be containerized in docker container, to do so you define dockerfile in your applications' code repository. When there is no event coming KEDA can scale in to zero instance and scale out to N instances based on the traffic and number of events. KEDA is supporting following Azure functions triggers:

- Azure Storage Queues.
- Azure Service Bus Queues.
- Azure Event / IoT Hubs.
- Apache Kafka.
- RabbitMQ Queue.

Serverless Platforms:

There is wide area of serverless services are available that you can use for your application. Serverless works like SaaS, you do not need to manage infrastructure, but unlike FaaS, some of the serverless require you to estimate your compute or data in the beginning of the subscription. Here is the list of mostly used serverless applications.

Google Cloud Serverless Options:

Compute:
    Cloud Functions.
    Cloud Run.
    App Engine.

Storage:
    Cloud Storage.

Cloud Firestore.

Messaging:
Cloud PubSub.
Cloud Tasks.

Data Analytics:
BugQuery.
Data Studio.

DevOps:
Cloud Build.
Cloud Scheduler.
Cloud IAM.
Cloud Logging.

Machine Learning:
AI Platform.
Cloud AutoML.

AWS Serverless options:

Compute:
AWS Lambda.
AWS Fargate.

Storage:
Amazon S3 (Simple Storage Service).
Amazon EFS (Elastic File System).
Amazon Dynamo DB.
Amazon Aurora Serverless.
Amazon RDS Proxy.

Messaging:

Amazon SNS (Simple Notification Service).
Amazon SQS (Message Queuing Service).

Analytics:
Amazon Kinesis.
Amazon Athena.

Azure Serverless options:

Compute:
Azure Functions.
Azure App Service.
Serverless Kubernetes using KEDA.

Storage:
Azure Blob Storage.
Azure Cosmos DB.
Azure SQL Database Serverless.

Messaging:
Azure Service Bus.
Azure Event Grid.
Cloud Events.

Analytics:
Azure Stream Analytics.
Microsoft Power BI.

DevOps:
Azure DevOps.
Azure Monitor.

Machine Learning:
Azure Machine Learning.

Azure Cognitive Service.
Azure Bot Service.

You can build end-to-end full stack serverless solutions using wide range of available serverless environments. You can focus on development and your business and get rid of infrastructure management. You can choose serverless solutions from your cloud provider or mix with third party service integration with your cloud service provider.

You can have combination of serverless services running on single cloud provider, combination of on-premises and cloud provider or combining multiple cloud provider. Let's say you are using Amazon Lambda function combining with Google cloud Storage or some third-party messaging service.

5

Platform as a Service (PaaS) Solutions

Platform as a service (PaaS) is a cloud computing execution model where cloud providers or third-party providers provide hardware and software platform and you will be responsible for your application and data.

How it is different than serverless? In PaaS you will have limited management responsibilities, you will define scaling criteria, set minimum and maximum number of running instances. At-least one instance will keep running in PaaS, number of instances will never go to zero, you will pay for the assigned resources. In serverless your instance may scale down to zero, you do not have to pay if there is no work for instances. In serverless, scaling will happen automatically as per load, you do not have to define minimum and maximum instance and setup scale based on uses, that provides true autoscaling capability.

Why PaaS:

Serverless compute option FaaS supports event driven system. FaaS is very expensive if your function is always running and number of instances almost never goes to zero. FaaS has some limitation for its capacity. You must have to think about supported events for FaaS and include that service into your design. You must have to check compatibility with your function before using any tool. In PaaS you will get little more autonomy than the serverless.

There are lots of PaaS providers available in the market, you should understand all of your requirements, what functionality and support you need in PaaS. All major cloud providers have PaaS platform for you. There are options from independent PaaS provider that you can use to deploy on any cloud provider and switch between cloud providers easily. If you will choose independent PaaS provider you can virtualized on-premises infrastructure and use it as PaaS, or you can virtualize your local infrastructure and use in combination of one or more public cloud providers as hybrid cloud.

I have seen several reports saying serverless is new PaaS, or Managed Kubernetes is new PaaS. This is not true. There are separate use cases for serverless, we already discussed those in above chapter, and there are separate use cases and organization architecture practices for using Managed Kubernetes, we will talk about Managed

Kubernetes in next chapter. PaaS is not going anywhere, even PaaS provider are increasing day by day, they are coming with new functionality and lot of supporting tools. I have seen two types of PaaS available in market, one type of platform supporting several languages runtime for deployment and running application over the PaaS platform. The other type of PaaS platform supports container runtime, where you can develop your application and containerize it to deploy on the PaaS platform. In both way you do not have to worry about the infrastructure and how your application is being managed, run, scaled, and deployed. But the flexibility with the container-based PaaS is that you can run your container anywhere, tomorrow you can move to Kubernetes cluster, on-premises, or Managed Kubernetes platforms. Container-based solution provide more freedom to move your application from one platform to another. In reality if you are not developing application for Serverless you can move to another platform anytime with very less effort.

There are many PaaS providers in market, some of the PaaS provider are very specialized in their area and target specific business functionality, for example Service-Now Platform provides capability to digitize workflow application quickly. Salesforce Platform provides capability to build Customer Relationship Management (CRM) application quickly and digitize CRM apps faster. Platform-as-a-Service (PaaS) sometimes referred as Application Platform as a Service (aPaaS).

AWS Elastic Beanstalk:

AWS Elastic Beanstalk is AWS Platform-as-a-service (PaaS) offering. You can deploy web application and services easily by simply uploading it. You can develop web application and services using Java, Node.js, .Net, Go, PHP, Python, and even you can deploy docker container on it.

There is no additional charge for using Beanstalk, you need to pay for all the assigned resources to store and run your application. You can ensure availability by deploying your application in multiple zone in Beanstalk. Elastic Beanstalk reduces management complexity without restricting control over application or choice of resources. Elastic Beanstalk automatically handles capacity provisioning, scaling, load balancing and health monitoring. You just need to provide the details in application configuration during deployment, for example you set maximum of 5 instance and minimum of 2 instance should be running in the configuration, Elastic Beanstalk will understand that when there is less load 2 instance of your application should be running, and it will increase the instance up to 5 based on increasing load on your application. Elastic Beanstalk application components are environments, versions, environment configurations, etc. Conceptually application is similar to folder in Elastic Beanstalk.

Application version in Elastic Beanstalk is label of deployable code that is unique for each deployable code and it distinguish one deployable code from another. You can deploy new version of code with a new unique label. You can go back to any of previously deployed version of your code, or you can deploy two version of code to measure the difference between two version of application code.

Environment is the collection of resources running an application version. One environment can run one application version at a time, but you can run different version of code in different environment. Based on configuration for the particular version of application you provided, Elastic Beanstalk will provide all the resource needed to run the application.

Environment configuration is the configuration that you provide with the application that determined your application behavior. You can refer save template of configuration, you can modify environment configuration anytime from console, CLI, or API, and you can save your environment configuration as a template.

Elastic Beanstalk provides variety of programming language runtime, application server, web server, and other third-party platforms to build your application. Once you develop application, create application configuration as per Beanstalk standard and then create bundle of the application that would be the unique version of application to deploy over Beanstalk platform. Once you deploy AWS Beanstalk will read the configuration and environment to run and decide the behavior of the application. After successful deployment, you can see the application metrics, events, and environment status in Elastic Beanstalk console, APIs, or AWS CLI (Command Line Interface).

Once you create the environment deploy the application, Elastic Beanstalk provisions all the required resources to run the application, like Amazon EC2 instances, auto scaling group, Elastic Load Balancer, etc.

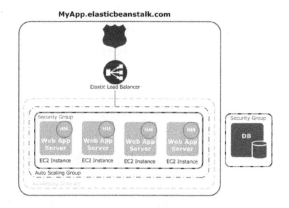

You can define security groups, firewall rules, ingress and egress rules, etc.

Here myapp.elasticbeanstalk.com url is having alias in Amazon Route 53, traffic from Amazon Route 53 goes to Elastic Load Balancer (elastic load balancing url like xxxx.us-east-1.elb.amazonaws.com). Amazon Route 53 is highly available and scalable DNS web service that provides secure and reliable routing.

The load balancer is the part of auto scaling group that distribute traffic to EC2 instances. Elastic load balancer automatically scale-out and scale-in instances based on the load and configuration provided for that application. The Elastic Load balancer always leaves minimum number of instance (at least one) in running status.

Based on container type elastic beanstalk decides technology stack and operating system. If it is .NET application using IIS container to run web server, elastic beanstalk will automatically choose windows server to deploy your application. Other than .NET (IIS server), elastic beanstalk chooses Linux servers to deploy the application.

Elastic Beanstalk deployment:

You have created one web application and planning to deploy on elastic beanstalk. You want to run multiple web servers in separate availability zones for high availability of the web application. You are planning to use MySql database that will run over Amazon RDS database instance. You are planning to put all the static resources on Amazon S3 bucket, Amazon CloudFront distribution for latency, Amazon ElasticCache cluster to run Memcached engine.

Below are the resources you have planned to use:

- Elastic Beanstalk platform for application deployment, run and server application.
- Amazon RDS for running MySQL instance.
- Amazon S3 to store static contents.
- Amazon CloudFront to deliver static content around the globe on request with reduced latency.
- Amazon ElasticCache's Memcached for database caching to improve performance and reduce load on the database.

Once you have planned you can simply follow few steps to deploy your application into elastic beanstalk.

- Create S3 Bucket, upload all the static contents into it.
- Use IAM user and policy to access contents from Amazon S3 bucket.
- Use elastic beanstalk console to upload source bundle and deploy it to your elastic beanstalk environment.

Elastic beanstalk provides several deployment policies and settings. You can choose right deployment policy for your application based on the nature of your application and business needs. Below are the policies supported by elastic beanstalk.

All at once, is the quickest deployment method. In this method your application will be unavailable for short period of time. All at once deployment policy deploy new version of application simultaneously on all the instances. While deployment occurs, application goes offline.

Rolling deployment is the solution to avoid downtime at a cost of longer deployment time. In this deployment elastic beanstalk splits the environments EC2 instances into batches and deploy the new version of the application one batch at a time. During deployment some instances servers request on old version of application while instances that completed batches serves other requests with new version of deployment application. After deployment in a batch, elastic beanstalk waits for all instances in that batch to become healthy before moving to another batch.

Immutable deployment policy ensures that deployment of new version of application happens on new instances instead of updating of the same existing instances.

Traffic-splitting deployment let you perform canary deployment, where you can split some percentage of the actual traffic to the new version of application and rest of the traffic will be routed to the old application. Elastic beanstalk launches new instances for this deployment like immutable deployment. If there is any issue in new deployment, new instances are not passing the health check, elastic beanstalk moves traffic to the old instances and terminates the new instances.

Google App Engine:

App Engine is fully managed PaaS provided by google. You can use App Engine to develop and hosting web applications at scale. It has support of lots of programming languages, libraries, and frameworks to develop your apps, App Engine will take care of provisioning of servers and scaling the instances based on load. App Engine is well suited for you if you are looking for PaaS and you are using microservice architecture for your application. App Engine comes with two flavor Standard Environment and Flexible Environment.

App Engine Standard Environment runs applications in a sandbox using runtime environment of supported languages. App Engine Standard Environment supports specific languages and specific version of language. Supported languages are Python 2.7, Python 3.7, Java 8, Java 11, Node.js 8, Node.js 10, PHP 5.5, PHP 7.2, PHP 7.3, PHP 7.4, Ruby 2.5, Go 1.11, Go 1.12, and Go 1.13. Standard Environment is like serverless as your instance count can go up to zero and you pay only for the period of time when you are using your resources.

App Engine Flexible Environment application runs within docker container on compute engine instances as node. You can run the code written in any programming language in docker container. App Engine supports popular development tools such as Eclipse, IntelliJ, Maven, Git, Jenkins, and PyCharm. You can use these development tools to develop, build, and deploy web and mobile application easily on App Engine.

Key feature of App Engine includes.

- It scans and detects common web application security vulnerabilities.
- It can instantly scale out and scale in your application automatically based on the load on your application.
- You can use Memcache in-memory data cache to improve application performance.
- You can use traffic splitting to route requests to the different app versions deployed, you can do A/B testing, and you can do incremental feature rollouts.

- You can use task queue that will allow application to perform work asynchronously outside of user requests. You can decouple application works using task queue.

App Engine Standard Environment:

App Engine Standard Environment is based on pre-configured container that runs on Google's infrastructure. App Engine is easy to deploy and reliable even in case of high load situations and dealing with very large amount of data. Your application will run in highly secure and scalable environment that is independent from hardware, operating system, or location of the server. The CPU and memory depend on the language runtime you use. Java runtime environment use more memory to run the application than any other environment. You can edit and override default instance class by using app.yaml file in your app directory.

The App Engine Standard Environment has two generations of runtime environment, First Generation and Second Generation. The Second Generation have improved capabilities and less limitations than the First Generation. Here is the difference between two generations.

First Generation	Second Generation
Supports Java 11, Python 3, Node.js, PHP 7, Go 1.12+ languages.	Supports Java 8, Python 2.7, PHP 5.5, Go 1.11 Languages.
Supports any language extension and system library.	Supports whitelisted extensions and libraries for Python 2.7 and PHP 5.5, and support any language extension and system library for java 8 and Go 1.11 languages.

App Engine Flexible Environment:

App Engine Flexible Environment provides flexibility of using either supported language runtimes or you can provide your own language runtime by supplying a custom docker image or dockerfile from the open source community. You can use custom libraries, debug your application using SSH, and deploy your own docker container if you are using All Engine Flexible Environment. You just provide hardware configuration for your application, like how much CPU and memory you need for this application and Flexible Environment will provide the infrastructure for your application based on your configuration.

App Engine Flexible Environment natively supports logging, versioning, security scanning, traffic splitting, authorization, SQL and NoSQL databases, and content delivery network (CDN).

App Engine Flexible Environment use managed virtual machines that ensures health check of instances, co-located instances with other services within the project for optimal performance that get handled by management service and heal your instance whenever needed. Critical updates for operating system will automatically applied to the operating system. VM will get restart every week, during restart any new operating system patches and security patches will get applied to the VM by management service. In flexible environment you can enable root access of your VM instances and access your VM using SSH. Flexible environment will run all long running instances, at least one instance will keep running even though there is no load on your application, it cannot be scale down to zero. As oppose to the standard environment flexible environment does not do billing by instance hours, flexible environment get billed based on the uses of CPU, memory, and persistence disk.

Azure App Service Environment (ASE):

Azure App Service is fully managed Platform-as-a-Service (PaaS) provided by Azure. You can develop and deploy web application and mobile backends with App Service easily and quickly. It is a http-based service for hosting web applications, REST APIs and mobile backends. App Engine supports .NET, .NET core, Java, Ruby, Node.js, PHP, and Python that can run on Linux or Windows based environments. App service automatically updates operating system and security patches on virtual machine. You can use DevOps tools like Azure DevOps, GitHub, Bitbucket, Docker Hub, or Azure Container registry. It will automatically scale-out or scale-in your application based on the load on your application, you can setup or manually do the scale out and scale in as well. It has in built with security features, you can authenticate user with Azure Active directory or federated identity (Google, Facebook, Twitter, and Microsoft). You can create IP address restrictions and manage service identities.

To create web application, App Service let you choose application template from Azure Marketplace, like WordPress, Joomla, etc. For most of the use cases App Service is the best choice for hosting web application in Azure cloud, but if your application is using microservice architecture, you can do the comparison with the Azure Service Fabric to choose the best option.

App Service environment is fully isolated and dedicated environment for running web applications. App Service is best if your application workload requires to run on very high scale. If you are migrating application to the cloud or building new application using microservice architecture, then App Service could be best choice for you.

You can create web app, mobile app, logic app, and API app using Azure App Service. You can use available third-party tools to integrate with your app, like WordPress, Joomla, Drupal, etc.

Azure Service Fabric:

Azure Service Fabric is the next generation Platform-as-a-Service (PaaS) from Azure that is way ahead of any App Service. App Service is useful for most of the web application, but Service Fabric is best choice for enterprise grade applications developed using microservice pattern.

Azure Service Fabric can run on any operating system whether it is windows or Linux, it can run on any cloud platform, you can run on on-premises, or even on your development machine. Service Fabric provides lightweight runtime that supports stateless and stateful microservices, that runs on cluster that is shared pool of virtual machines. Service Fabric hosts microservices inside containers, and those containers are deployed across Service Fabric cluster. Service Fabric is Microsoft's container orchestrator that deploy microservices across the cluster of machines.

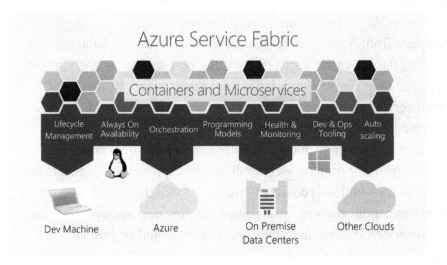

With Service Fabric, you can:
- Deploy to Azure or to on-premises datacenters that run Windows or Linux with zero code changes. Write once, and then deploy anywhere to any Service Fabric cluster.
- Develop scalable applications that are composed of microservices by using the Service Fabric programming models, containers, or any code.

- Develop highly reliable stateless and stateful microservices. Simplify the design of your application by using stateful microservices.

- Use the novel Reliable Actors programming model to create cloud objects with self-contained code and state.

- Deploy and orchestrate containers that include Windows containers and Linux containers. Service Fabric is a data aware, stateful, container orchestrator.

- Deploy applications in seconds, at high density with hundreds or thousands of applications or containers per machine.

- Deploy different versions of the same application side by side and upgrade each application independently.

- Manage the lifecycle of your applications without any downtime, including breaking and nonbreaking upgrades.

- Scale out or scale in the number of nodes in a cluster. As you scale nodes, your applications automatically scale.

- Monitor and diagnose the health of your applications and set policies for performing automatic repairs.

- Watch the resource balancer orchestrate the redistribution of applications across the cluster. Service Fabric recovers from failures and optimizes the distribution of load based on available resources.

You can deploy Service Fabric application using multiple ways, but Microsoft suggests using Azure Resource Manager (ARM) to deploy application. In ARM, you can describe application and services in JSON, and then deploy them in the same ARM template as your cluster. ARM template is the JSON file that defines infrastructure and configuration of the projects, you can create ARM template once and deploy multiple times using same template that provides consistency in deployment.

Cloud Foundry (CF):

Cloud Foundry is open source multi cloud Platform-as-a-Service (PaaS). You can use any cloud provider that provides IaaS to deploy, manage, and scale your application with few configurations. Cloud Foundry is one of the fastest ways to develop and deploy application on multi-cloud and hybrid-cloud platform. You can switch cloud provider with less effort if you are using Cloud Foundry for PaaS. Cloud Foundry uses BOSH, an open source tool, for configure, deploy, manage, scale, and upgrade on any cloud IaaS provider.

BOSH is an open source project that can provision and deploy software over large number of virtual machines and perform monitoring, resiliency, and software updates with near zero downtime. BOSH supports large distributed systems and many IaaS (Infrastructure-as-a-Service) providers includes, VMware, Google Cloud Platform (GCP), Microsoft Azure, Open Stack, Amazon Web Services (AWS), Alibaba Cloud, Apache CloudStack, and VirtualBox. Using BOSH you can create and deploy virtual machines on top of physical infrastructure, and run the Cloud Foundry on top of this cloud.

One advantage of using Cloud Foundry is to get freedom of using any programming language to develop your application and use any cloud. If you use PaaS directly from cloud provider, you may have restrictions of developing application in provider supported programming languages.

Cloud Foundry cloud controller is used to manage app lifecycle, runs the application and other process on virtual machines, and balance the demand based on load.

If load balancer is provided by customer, Cloud Foundry Gorouter works with load balancer to route the traffic based on the demand.

If you think from development prospective, few cf command do the magic in Cloud Foundry. Developers do not have to understand the internal architecture of infrastructure where application runs. Developers and DevOps team will be able to push the application with just cf push command, does not matter if he is pushing his application to the Google Cloud, Azure, AWS, or on-premises private cloud.

Cloud Foundry includes Stack and Buildpack with the application source code. Stack is the operating system your application runs on. Buildpack contains all languages, libraries, and services that application can use at the time of deployment.

Cloud Foundry use Diego to manage application containers, that is a self-healing system that try to keep the correct number of instances running in Diego cell. Diego schedules and run tasks and long running process.

The following diagram provided by Cloud Foundry that illustrates steps and components involves in the process of staging a Buildpack application.

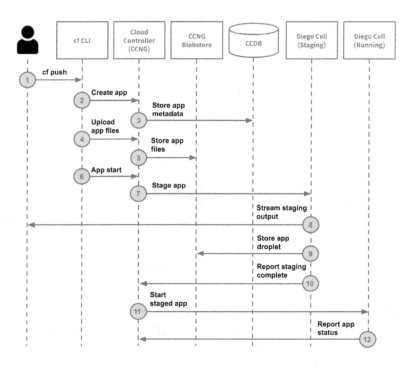

The above diagrams show how Diego stage application using Buildpack. You can stage docker image on Cloud Foundry instead of using Buildpack.

Cloud Foundry provides following diagram that illustrates the steps and components involves in the process of staging a Docker image using Diego.

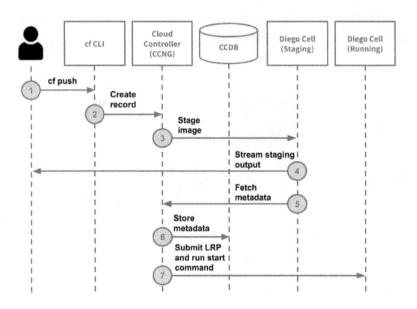

You can use Cloud Foundry if you are planning to use multi cloud or Hybrid Cloud. You can use docker container to deploy on Cloud Foundry that will provide you flexibility to move freely with any vendor.

If you are planning to use Kubernetes but likes the features of Cloud Foundry, you can explore KubeCF, that is Cloud Foundry Application Runtime for Kubernetes.

6

Infrastructure as a Service (IaaS) Solutions

Infrastructure-as-a-Service (IaaS) is a service model, that delivers compute, network, and storage resources on demand over the internet on pay-as-you-go basis. IaaS is like your data center, instead of having your data center or instead of provisioning more servers in your data center you are provisioning data center from cloud provider's data center and save time of procurement.

IaaS provides some flexibility as you can immediately provision the server whenever needed. It will take lot of time to purchase new server for your data center, so you need to plan early enough so that you will have enough server to scale your application properly. Usually you keep some buffer server for peak load time, and sometimes you use your lower environment server as production server till the time you get new server.

In some situation, let's say some retailers keep pool of extra servers for peak period like thanksgiving time, and most of the time in year those server remains sitting ideal. If pool of servers is sitting ideal most of the time in year does not mean there is no cost associated on those servers during those ideal time, you need to keep your server updated, install all security and operating system patches even during ideal time. In some year you may need to upgrade your server as well as new processor comes in market with better efficiency.

With IaaS on cloud you can save these extra costs, you can provision all the servers and network based on your need and you will pay only for those resources you are provisioning. For example, during thanksgiving you can provision more server by scaling up your system and when load is less you can decommission your servers by scaling down the system. You will be paying for extra server only during thanksgiving and you do not need to worry about keeping and maintaining extra server like you do in your data center.

Other saving is over depreciation of your computing resources. Servers on your data center will lose their value over time, but when you will use IaaS, you do not have to worry about depreciation cost on the server, it is cloud provider's job to upgrade their system. You can upgrade your server any time with just call of an API, or by executing a command, or by using cloud providers console in a minute, but upgrading the server, server's memory on your data center will take time depends on buying process of your organization.

Another benefit of IaaS is geographical presence of data center. If your business is growing over multiple states or countries, it will take time to setup data center in different region. You need to rent or buy building for data center, you need to ensure 24X7 electricity supply, and better high-speed network connectivity. All these processes will take lots of time to setup the data center that may delay your business in new areas.

If your business is very time sensitive and delay in launch of your service may lose the edge of your business, then IaaS is better option than the setting up your own data center. If you just calculate the cost of data center vs IaaS, then data center may look cheaper, but if you will calculate the total operating cost, depreciation, and opportunity cost then IaaS will be cheaper and future safe. In IaaS, you do not need to invest upfront, you can provision server for short period of time for your research and development work, you can do POC, or your can try out the market by provisioning the server and continue only if your business is successful.

Amazon EC2:

Amazon Compute Cloud (Amazon EC2) is a compute service that provides servers in Amazon's data center. Amazon EC2 provide range of options of virtual machines starting from entry-level economical VMs for dev/test to general purpose VMs, burstable VMS, memory optimize VMs, compute optimize VMs, storage optimized VMs, and GPUs. You can use Amazon EC2 to launch as many servers as you need in the choice of your Amazon's data center, Amazon's data centers are available at multiple location worldwide. Once you will provision EC2 instances, you can configure security and networking and manage storage. You will be able to scale out and scale in your servers horizontally based on the load of your application.

Amazon EC2 provides the following features (refer AWS documentation for details):
- Virtual computing environments, known as instances.
- Preconfigured templates for your instances, known as Amazon Machine Images (AMIs), that package the bits you need for

your server (including the operating system and additional software).

- Various configurations of CPU, memory, storage, and networking capacity for your instances, known as instance types.
- Secure login information for your instances using key pairs (AWS stores the public key, and you store the private key in a secure place).
- Storage volumes for temporary data that's deleted when you stop or terminate your instance, known as instance store volumes.
- Persistent storage volumes for your data using Amazon Elastic Block Store (Amazon EBS), known as Amazon EBS volumes.
- Multiple physical locations for your resources, such as instances and Amazon EBS volumes, known as Regions and Availability Zones.
- A firewall that enables you to specify the protocols, ports, and source IP ranges that can reach your instances using security groups.
- Static IPv4 addresses for dynamic cloud computing, known as Elastic IP addresses.
- Metadata, known as tags, that you can create and assign to your Amazon EC2 resources.
- Virtual networks you can create that are logically isolated from the rest of the AWS cloud, and that you can optionally connect to your own network, known as virtual private clouds (VPCs).

To create EC2 instance in AWS, you can choose a virtual machine from large set of options available. Options ranging from micro virtual machine to high memory, or high compute options. You need to create key pair for your virtual machine and then create security group for EC2 instances. Security group is virtual firewall that is similar to physical firewall but have lot more feature and flexibility than the physical firewall. When you launch EC2 instance you need to mention one or more security groups, if you do not provide any security group EC2 will use default security group. Default security group will allow all outbound traffic but deny all inbound traffic. AWS security groups are always permissive, you can set rule to allow certain access, you cannot create rule to deny certain access. As by default all access are denied, you just need to create rule to allow access.

You need to create VPC, VPC is virtual private cloud that enable you to launch AWS resources into a virtual private network that you have defined. Virtual network is similar to the physical network that you define in your data center. Virtual network allows you to create scalable infrastructure of AWS. Once you create VPC, you can create subnet with the range of IP address in the VPC, create router table to determine how network traffic is directed, internet gateway that can be attached to the VPC, VPC endpoint to privately connect your VPC.

Once you define your network, you can create EC2 instance in your VPC. You can decide type of storage attached to your EC2 instance. Main advantage with EC2 instance over local data center is, you can auto scale your EC2 instances. You can create auto scaling group and specify minimum and maximum number of instance in the auto scaling group, EC2 auto scaling ensure that your instance will not go below to the minimum number of instance defined and not go beyond the maximum number of instance defined for maximum number of instance.

You can create instance configuration template that auto scaling group can use to launch EC2 instance with the launch configuration. Launch configuration will have information about instances to be launched, like ID of the Amazon Machine Image (AMI), the instance type, a key pair, one or more security group, and block device mapping with the instance. You can specify one launch configuration for one auto scaling group at a time, but you can use same launch configuration for multiple auto instance group. Once you launch the instance, you cannot change launch configuration.

Azure Virtual Machines (Azure VMs):

You can create Linux and Windows virtual machine in seconds using Azure VMs. Azure VMs provide range of options of virtual machines starting from entry-level economical VMs for dev/test to general purpose VMs, burstable VMS, memory optimize VMs, compute optimize VMs, storage optimized VMs, and GPUs. You can select from wide range of options from available range of option according to the requirement and workload you have.

Azure provide different solutions and images for its VMs, you can have your own image or third-party image for your VMs. Once you can create VPC and networking to isolate your VMs, implement firewall, and decides how to route traffic.

Azure Dedicated Hosts:

Azure Dedicated Hosts is a service that provides physical servers on that you can host one or more virtual machines; these dedicated hosts are tied up with one Azure subscription. Dedicated hosts are similar to the servers in data center, provided as a resource, you can provision dedicated host within a region, availability zone, and fault domain. You will not be able to scale dedicated hosts; you can use this machine as standalone dedicated servers.

Dedicated host having following benefits:

- You will have reserved dedicated host where only your virtual machine will run. Hardware separation of dedicated host will be at physical server level. Your dedicated host will be deployed on Azure's data center and it can share some network and underlying storage infrastructure. You can think of workload run on the server which is very sensitive, and you want them to run on dedicated server like in your data center.
- Majority of maintenance event will have very little or no impact on your virtual machines on dedicated servers. You can opt-in for maintenance window to reduce impact on

your service. You can run such workload on dedicated server whose one second on downtime may have significant impact.

- You can get Azure Hybrid benefit if you have windows license or SQL server license and you want dedicated server for windows or SQL server. You can use the same license in dedicated server. However, you can use this hybrid benefits even if you are using virtual machine in shared environment.

Azure VMware Solution (AVS):

Azure VMware solution (AVS) provides you private cloud in Azure, contains vSphere clusters, built from dedicated bare-metal Azure infrastructure. This private cloud will be provisioned with vCenter server, VSAN, vSphere, and NSX-T. Your private cloud host can be scaled from 3 to 16 hosts, and you can have multiple cluster in single private cloud.

It is VMware validated solution with ongoing validation and testing of enhancements and upgrades. The private cloud infrastructure will be maintained by Microsoft, so that you can focus on development and running of workload on the private cloud. Below is the diagram provided by Azure explaining AVS.

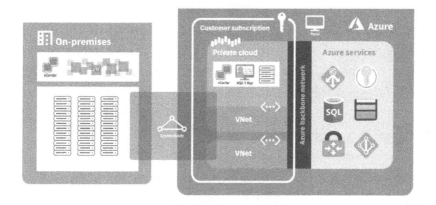

Spot Virtual Machine (Spot VM):

Spot VM allow you to use unused pool of VM from Azure at very low cost. At any point of time when Azure needs it VM back it can take it from you. If your workload can handle interruptions like batch processing job, large computing workloads, etc. then you can use spot VM.

You can opt-in to receive in-VM notification before your spot VM will get eviction. You will have only 30 second of time to run any shutdown script you already configured for the same purpose, so that you can finish your work, perform shutdown task, or do some cleanup work before eviction of VM.

There are few limitations for spot VM, you cannot get B-series, or promo version of VMs. Spot VMs cannot use ephemeral OS disks. Spot VMs can be deployed to any region except Azure china 21Vianet.

Once created, spot VM will be working like standard VM, but there is no SLA provided for spot VMs.

SAP HANA on Azure (Large Instances):

SAP HANA on Azure is a unique solution to Azure that provides virtual machines for deploying and running SAP HANA. This solution is builds on non-shared host or server on bare metal hardware just assigned to you.

It does network isolation to isolate customer within infrastructure stack through virtual networks per customer assigned tenants. The network isolation between tenants prohibits network communication between tenants in the infrastructure stamp level. There will be storage isolation by assigning separate storage per tenant. Server or host will also have isolation and cannot shared between customer or tenants.

SQL server on Azure:

SQL server on Azure allow you to manage full version of SQL server on dedicated server in Azure, without having to manage any on-premises hardware. You can have multiple license options even for pay-as-you-go licenses. You can provision SQL server on Azure in any region around of world and from available multiple VM size, based on your need.

You can take advantage of automatic patching to schedule a maintenance window, or you can have advantage of automatic backup which can take regular backup of your database to the blob storage. You can also configure SQL server availability group for high availability. If you already have SQL server in your on-premises, you can save money by bring your own license (BYOL) after moving your current SQL server to the Azure cloud.

Virtual Machine Scale Set:

Azure virtual machine scale set will allow you to create group of identical load balanced virtual machines that you can manage easily. Number of virtual machines will automatically decrease or increase as defined based on the load.

Scale set will provide redundancy the workload on multiple virtual machine, that provides you high availability, and you can access your application through load balancer. Scale set is being used to run multiple virtual machines, if one of the virtual machines has an issue, then your will continue access your application from one of the other virtual machines in scale set.

Virtual machine scale set provides high availability and resiliency. Scale set supports up to 1000 virtual machine instances if you are creating virtual machine using provided template, if you are creating your own custom virtual machine instances then this limit will be 600 virtual machine instances.

You can enable monitoring for virtual machine scale set with Azure Application Insights to collect detail information about application including exceptions occurred in application. You can use Azure Monitor for VMs, which will automate the collection of important CPU, memory, disk, and network performance counters form the VMs in your scale sets. It also includes additional monitoring capabilities that helps you to focus on availability and performance of your scale sets.

Google Compute Engine:

Google Compute Engine is an Infrastructure-as-a-Service (IaaS) where you can provision virtual machine instance hosted in Google Cloud's infrastructure. You can create new virtual machine instance by using Google Cloud Console, the gcloud command-line tool, or the Compute Engine API. Compute Engine instances can run the public image for Linux or Windows server, as well as private image that you can create or import from existing system. You can also deploy docker image containers that will run container-optimized operating system public image.

Google cloud comes with the concept of project. The top level of hierarchy is organization, that is optional and not available if your subscription is free trial subscription. Google Cloud resource hierarchy helps you to manage resources. Google Cloud resource hierarchy provide a hierarchy of ownership, which binds the lifecycle of resource to its immediate parent in the hierarchy. It also provides inheritance for access control and organization policies.

Organization is the top level of the hierarchy that does not have parent, all resources of organization will have exactly one parent.

Folders are an additional grouping mechanism on top of projects. Folders and projects are mapped under organization resource. You need to have organization to create folders. Both cloud IAM and organization policies are inherited though the hierarchy. Below diagram represents Google Cloud resource hierarchy.

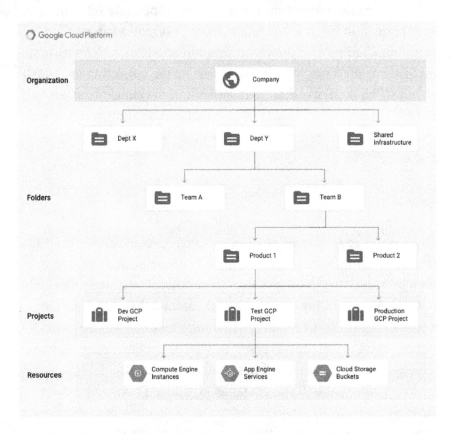

Each Google Compute Engine instance comes with a small boot persistence disk that contains operating system. You can add additional storage option to your instance based on the need.

In one project you can have up to five Virtual Private Cloud (VPC) network. Each Compute Engine instance belongs to one VPC network. Compute instances in the same VPC network can communicate each other using Local Area Network (LAN) protocol.

A VPC network is a virtual version of physical data center network, that provides connectivity between your cloud resources in a project. Cloud resources could be Compute Engine VM instances, GKE clusters, App Engine Flexible environment instances, or any other resources. A new project starts with a default VPC network having one subnetwork (subnet) in each region. VPC networks, including their associated routes and firewall rules are global resources in GCP, they are not associated with any particular resign or zone. Subnets are reginal resources; each subnet defines a range of IP address. If network firewall allows, resources within a VPC network can communicate with each other by using private IP address. To keep VPC network in a common host project, you can use shared VPC. All projects should belong to same organization for shared VPC. A VPC network can be connected to the other VPC network in different projects or organization using VPC Network Peering. Cloud VPN or Cloud Interconnect can be used to securely connect VPC network in Hybrid environment.

You can create VM from a public image or a custom image. In the process of creating a Compute Engine instance, you need to select a zone where you want to create your VM, select a network, and select a subnet. Google cloud assign IP address to the instance from the range of available IP address in the subnet.

Like AWS and Azure you can create Preemptive virtual machine instance that cost much lower than normal instance. You can use preemptive can be terminated any time and will get terminated in 24 hours. Preemptive instances cannot live migrate to a regular VM instance and are not covered by Compute Engine service level agreement (SLA). Preemptive instances send preemption notice to the instance and terminated after 30 seconds after sending signal. You can provide shutdown script at the time of creating preemptive instances, so that you can get completion status of running task and do the cleanup.

Other IaaS providers:

There are many IaaS provider other than AWS, Azure, and Google that may cost less and may be better choice for your requirement. Below are few names of IaaS providers.

- Rackspace Open Cloud provides public, private, and multi-cloud.
- DigitalOcean provides cloud computing solution designed for developers.
- HP Enterprise ConvergedSystem provides Public, Private, and Hybrid cloud.
- IBM Cloud provides complete cloud solution with Public, Private, and Hybrid cloud.
- OpenStack provides IaaS in both Public and Private cloud.
- There are many more IaaS providers available.

7

Managed Kubernetes Solutions

Kubernetes is an open source container orchestration engine that manage containerized workload and service. Here container could be any container but mostly referred to a Docker container.

Container is a logical packaging mechanism where application is abstracted from the environment in which they run. Initially developed by google and open sourced in 2014.

In traditional deployment, application runs on physical servers. If multiple applications are running in one physical server, they all share the same resources and there is no way to do the separation. If one application consuming lot of memory other application might get impacted.

Then virtualization was introduced. In virtualization, you can run multiple virtual machine on a physical server, each virtual machine will run application, each application will get their separate space, isolated from each other. Each virtual machine works like a physical machine with its own operating system.

Then it comes to container. Container is like virtual machine runs in the isolation with each other but share the same operating system. Because of this property containers are lightweight and similar to the VMs.

Kubernetes takes care of scaling and failover. Below is the detail of feature provided by Kubernetes.

- Service discovery and load balancing: Kubernetes can load balance and distribute the traffic, also can expose a container using its DNS name or using their own IP address.
- Storage Orchestration: Kubernetes can mount local storage, public cloud storage, or any other storage automatically.
- Automated Rollouts and Rollbacks: You can automate Kubernetes to create new containers for your deployment, remove existing container and adopt all the resources to the new containers.
- Automatic bin packing: You can mention how much CPU and RAM each container requires. Kubernetes fit the container in appropriate nodes to make the best use of the resources.

- Self-healing: Kubernetes restarts containers that fail, replaces containers, kills containers that doesn't respond to your user defined health check, and doesn't advertise them to the client until they are ready to serve.
- Secret and Configuration management: Kubernetes let you deploy and update secret and application configuration without rebuilding your container images, and without exposing your secret in your stack configuration.

Kubernetes Components:

A Kubernetes cluster consists of a set of worker machines, called nodes that runs containerized application. Every cluster has at least one node. Components of application workload resides in Pods. A Pod can have minimum one container that will be hosted by worker node. Below are the components of Kubernetes:

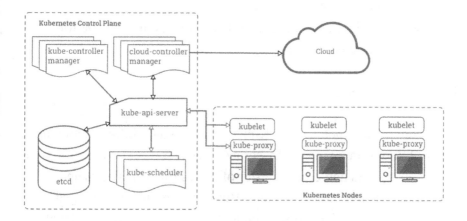

- Kube-apiserver: is the front end of the Kubernetes control pane that expose the Kubernetes API. You can run multiple instance of apiserver and balance traffic between instances.
- Etcd: is a consistent and highly available key value store used as Kubernetes backing store for all cluster data.

- Kube-scheduler: watches for newly created Pods which is not assigned to any node, select a node for them to run on.
- Kube-controller-manager: runs controller process that is responsible for noticing and responding when nodes goes down, maintaining the correct number of Pods, populates the endpoint objects, and create default account and API access token for new namespaces.
- Cloud-controller-manager: link your cluster into your cloud provider API, it only runs those controllers that are specific to your cloud provider.
- Kubelet: is an agent that runs on each node in the cluster and make sure that containers are running in a Pod. It manages only those containers created by Kubernetes.
- Kube-proxy: is a network proxy that runs on each nodes of cluster and maintains network rules that allow network communication to your Pod from network sessions inside or outside of your cluster.
- Container runtime: is a software that is responsible for running containers.

Managed Kubernetes:

Managed Kubernetes is the Kubernetes in which infrastructure and cluster of nodes are managed by someone else. You can virtualize your data center servers and create cluster of nodes for Kubernetes and allocate them resources on demand. But if you are using Managed Kubernetes, you can have operation and development team to develop and deploy on the Kubernetes cluster. You will not worry about nodes and resource availability in Kubernetes cluster.

Managed Kubernetes is managed service, in which cloud provider will create clusters, monitor them and manage those clusters. Managed Kubernetes is gaining popularity very fast. Managed Kubernetes let you run Kubernetes on Public, Private, and Hybrid clouds in automated and API driven infrastructure. Almost all cloud providers providing Managed Kubernetes services. There are lots of managed private Kubernetes cluster tools are available that can be used to manage Kubernetes cluster on on-premises servers.

Amazon Elastic Kubernetes Service (Amazon EKS):

Amazon EKS is fully Managed Kubernetes service that provides scalable and highly available control plane that runs across multiple availability zones to eliminate a single point of failure. It automatically detects and replace unhealthy control plane nodes, and provides on-demand, zero downtime upgrade and patching.

With Amazon EKS, you can use AWS Fargate which is a serverless compute for controllers. With Fargate you do not need to provision and manage servers. Fargate choose right amount of compute, eliminating the need to choose instances and scale cluster capacity. You only need to pay for the resources required to run your cluster.

Amazon EKS is integrated with AWS services makes it easy to monitor, scale, and load balance your application. EKS automatically applies latest security patch to your cluster control plane. Below figure from AWS showing the working flow of EKS.

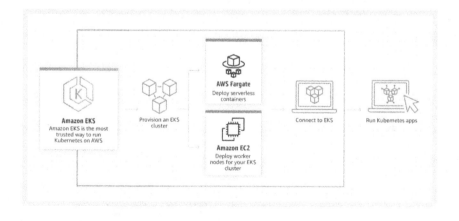

Amazon EKS can be used on AWS Outposts to run containerized application to on-premises systems. AWS Outposts is a fully managed service that extends AWS infrastructure, AWS services, APIs, and other tools to virtually any data center, co-location space, or on-premises facility. With AWS Outposts, you will be able to manage containers in the same way you manage containers in the cloud. Outposts are connected to the nearest AWS region to provide the same management and control plane service on-premises for a fully consistent operation experience across on-premises and cloud environment. AWS Outposts infrastructure and AWS services are managed, monitored, and updated by AWS just like the cloud.

You can model your machine learning workflows using Kubeflow with EKS. To run training and inference in TensorFlow on EKS you can use AWS Deep Learning Containers which allow you to setup deep learning environment with optimized, pre-packaged container images. It supports TensorFlow, PyTorch, and Apache MxNet.

You can build web applications that automatically scale up, scale down, and run in highly available configuration across multiple availability zones. You will get performance, reliability, and availability for your we application deployed on EKS, also you will get flexibility of container.

Sequential and parallel batch workloads can be deployed on EKS cluster using the Kubernetes Jobs API. Using EKS, you can plan, execute, and schedule your batch computing workloads across AWS Compute services like Amazon EC2, Fargate, and Spot Instances.

Azure Kubernetes Service (AKS):

Azure Kubernetes Service (AKS) is a Managed Kubernetes service that make it simple to deploy your workload container on Managed Kubernetes cluster in Azure. It reduces the complexity and operational overhead of managing Kubernetes by offloading much of the responsibilities to Azure. Azure handles critical tasks such as health monitoring and maintenance for you. As a Managed Kubernetes service, AKS is free, you will only pay for agent nodes within your clusters, not for master nodes.

A Kubernetes cluster is divided into two components, control plane nodes which provides the Kubernetes services and orchestration workloads, and the nodes that runs your application workloads. This is described by Azure in following diagram.

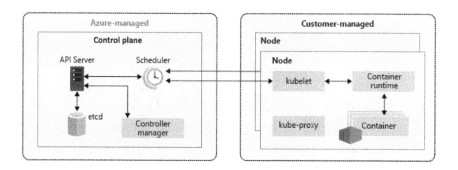

Control Plane:

Control plane is provided as a managed Azure resource abstracted from the user. There is no cost associated with control plane.

The control plane has API server, that is how underlying Kubernetes APIs are exposed. Etcd is a key-value store within Kubernetes that maintains the state of Kubernetes cluster and configuration. Scheduler schedule the workload over the node to run. Controller manager handles node operation and replication of Pods.

AKS provides single-tenant control plane, with a dedicated API server, scheduler, etc. You define number and size of node and Azure configures secure communication between control plane and nodes. Control plane is fully managed that means you don't need to configure components like highly available etcd store, and you can't access control plane directly. If you want to configure control plane is a specific way or you want to access the control plane, you can deploy your own Kubernetes cluster using aks-engine.

Customer managed Nodes:

An AKS cluster can have one or more than one node which is an Azure virtual machine that runs the Kubernetes node components and container runtime. Azure has below diagram to explain the node created out of Azure VM.

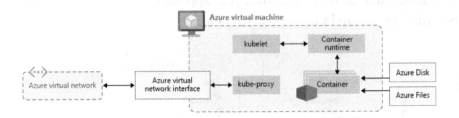

The Kubelet is the Kubernetes agent that processes the orchestration requests from the control plane and scheduling of running the requested containers. On each node virtual networking is handled by kube-proxy that routes network traffic and manages IP addressing for services and Pods. The container runtime is the components that allows containerized application to run and interact with the additional resources such as virtual network and storage. AKS uses Moby as the container runtime.

In AKS, you can create cluster using the virtual machine with Ubuntu Linux and Windows server 2019 image. When you create AKS cluster, or when you scale out the existing cluster, Azure creates the requested number of VMs and configure them automatically, you do not need to do the manual configuration.

Nodes of the same configuration are grouped together called node pools. A Kubernetes cluster can have one or more than one node pools. To ensure your cluster operates reliably, you should have at least two nodes in the default node pool.

If there are many node pools available in AKS cluster, you may need to tell Kubernetes scheduler that which node pool to use for a given resource. NodeSelector does this job, you provide value to NodeSelecter to tell the scheduler which node pool you want to select.

AKS manage applications in Kubernetes cluster using Helm. To use Helm, a server component Tiller should be installed into your

Kubernetes cluster. The Tiller manages the installation of chart in the cluster. The Helm client should be installed locally into your computer, or you can use it from the Azure Cloud Shell. Below diagram describe the process.

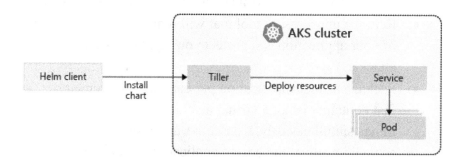

Google Kubernetes Engine (GKE):

GKE is Managed Kubernetes engine with auto-scaling and multi-cluster support. Kubernetes was developed initially by Google and then Google made it open source. Google claims that they know best how to manage and run Kubernetes cluster for your application.

When you are running GKE you are benefiting from below mentioned Google's advance cluster management features.

- You will use Google cloud load balancing for Compute Engine instance.
- Node pools to designate subsets of nodes within a cluster of additional flexibility.
- Automatic scaling of cluster's node instance count.
- Automatic upgrade of your cluster's node software.
- Node auto-repair to maintain node health and availability.
- Logging and monitoring with Google cloud's operations suite for visibility into your cluster.

You can follow best practices for GKE that is similar to the 12-factor application. Your workload will be wrapped in the container and container will be in Pod. Pod will get scheduled in Kubernetes cluster. How you are containerizing your application is very important. You can follow below best practices while containerize your application.

- Remove unnecessary tool that will reduce the attack surface of your application and protect your applications from attacker.

- Package a single application per container. Be remember that container is not a virtual machine that can run different things simultaneously. Containers are designed to have the same lifecycle as the hosted application inside it, your container should have only one application. Container and application inside it should start and stop at the same time. If you start the container your app should start and when you stop your app, your container should stop as well. If you have multiple app in a container then they might have different lifecycle or different states. There would be a possibility that your container might be running but its core application is crashed or unresponsive.

- Optimize for Docker build cache. The Docker build cache can accelerate the building of container image.

- Build the smallest image possible. Smaller image can be downloaded or upload faster by node.

- Properly tag your image. Docker image generally identified by their name and their tag. You can tag in source code repository like git with the version number with the git tag command.

GKE containers and workloads can be easily integrated with all google cloud services and offerings. Using Google cloud continuous integration and continuous delivery tool you can easily build and serve application containers. You can use Cloud Build to build Docker image (or any other container image) from any of the supported source repository like git. You can use Container Registry to store and server container image.

Other Platforms:

There are many other platforms that provides manages Kubernetes service and may be worth looking for. Below are the few examples.

- Rackspace Kubernetes-as-a-Service provides fully Managed Kubernetes service that helps you to improve application portability across cloud and internal environments.
- Platform9 Managed Kubernetes (PMK): It is a fully automated fully Managed Kubernetes service that you can use in data center or public cloud environments.
- DigitalOcean Managed Kubernetes solution.
- VMware Cloud PKS: Jointly developed by VMware and Pivotal. VMware Cloud PKS is multi-cloud ready with initial support with AWS.

8

Database Solutions

Deciding database is one of the most critical decision usually architects take. When deciding for the database options you should know the amount of data you are going to store, amount of consistency you need, and pattern of store the records, and search of records by application.

Historically Database was optimized based on consistency and ability to store and search the data faster based on indexing. To achieve this, we were using RDBMS (Relational Database Management System) that is the DBMS (Database Management System) stores data in structured format, using rows and columns called Database Table. You store related data distributed in multiple table and make a relationship between those tables. To query the data from RDBMS we use SQL (Structured Query Language). Later NoSQL databases developed. NoSQL databases can be document based, key-value pairs, or Graph databases. The biggest advantage of NoSQL databases are they can scale horizontally, the limitation of SQL databases are they can scale only vertically. Some of the SQL database came later that can scale horizontally like Google Cloud Spanner.

We will go through all the datasets provided by cloud providers with its feature that will help you decide the database for your application. You can do more research and comparison by researching more in depth.

Databases in AWS:

AWS provides about 15 database engine including relational database, key-value database, document database, in-memory database, graph database, time series database, and ledger databases.

AWS relational Databases:

Relational databases might be most suitable for traditional applications, ERP, CRM, and e-commerce applications. You can always do mix of relational database with NoSQL databases. Amazon Arora, Amazon RDS, and Amazon Redshift are the relational database offering from AWS.

Amazon Arora is compatible with MySQL, and PostgreSQL and is five times faster than MySQL and three times faster than standard PostgreSQL. Arora is distributed, fault-tolerant, self-healing, storage system that auto scales up to 64 TB per database instance. It delivers high performance and availability with up to 15 low latency replicas. It provides point-in-time recovery, continuous backup to Amazon S3, and replication across three availability zones. You can migrate your MySQL and PostgreSQL with AWS Database Migration Service for a secure migration with minimal downtime. A single Arora database can span multiple AWS regions to enable fast local reads and quick disaster recovery. Arora is fully managed, meaning you do not need to worry about hardware provisioning, software patching, setup, configuration, or backups. You can monitor database performance using Amazon CloudWatch. Arora provides encryption of data at rest using key you create and control through AWS KMS and it also provides encryption of data in transit using SSL.

Amazon Relational Database Service (Amazon RDS) is a relational database in AWS cloud that provides quick setup. It provides cost efficient and resizable capacity, automate administration tasks for you, like hardware provisioning, database setup, patching, and backups. Amazon RDS allows you to choose from six database engines, Amazon Arora, PostgreSQL, MySQL, MariaDB, Oracle Database, and SQL Server. Using AWS Database Migration Service, you can easily migration your database to the Amazon RDS or replicate your existing database to Amazon RDS. Amazon RDS is less expensive and you have to pay for the instance you configure for RDS. You can choose SSD backed storage option to make RDS faster. If you will compare RDS with Arora, Arora is fully managed and can scale to 6 availability zone, you will have less operating cost if you are choosing Arora. RDS may be faster if you will choose right instance type and storage for it.

Amazon Redshift database is an enterprise level, petabyte scale, fully managed data warehousing service. Amazon Redshift is based on PostgreSQL 8.0.2. Amazon Redshift is specially designed for online analytics processing (OLAP) and business intelligence (BI) applications, which requires complex queries against large datasets. The specialized data storage schema and query execution engine that

Amazon Redshift uses are completely different from the PostgreSQL implementation. Amazon Redshift stores data in columns. Amazon Redshift provides efficient storage and optimum query performance through a combination of massively parallel processing, columnar data storage, and very efficient targeted data compression encoding schemas. Amazon Redshift creates one database when you provision a cluster. This is the database you use to load data and run queries on your data. You can create additional database as per your need by running a SQL command. Using Redshift, you can query petabyte of structure and semi structure data across your data warehouse, operational database, and your data lake using standard SQL. Amazon Redshift is comparable with Google Cloud Big Query.

AWS NoSQL Databases:

Amazon Dynamo DB is fast flexible NoSQL key-value store suitable for high traffic web applications, e-commerce systems, and gaming applications. DynamoDB delivers single digit millisecond performance at any scale. DynamoDB global tables replicate your data across multiple AWS regions to give you fast and local access to data for your globally access application. If your use case requires even faster access with microsecond latency, you can use DynamoDB Accelerator (DAX) that provides fully managed in-memory cache. DynamoDB is a serverless database, meaning, you do not have to provision any server, you do not need to install any software, no maintenance is required, and there is no operation cost associated with this database. DynamoDB is available in two modes, provisioned and on-demand capacity modes, you can optimize cost by specifying capacity per workload, or paying for only the resources you consume. DynamoDB supports ACID transactions, it encrypts all the data by default. You can create full backups of terabytes of data instantly with no performance impact to your tables and recover to any point in time in the preceding 35 days with no downtime.

Amazon DocumentDB is fast, scalable, and highly available MongoDB-compatible database service. You can easily store, query, and index JSON data. DocumentDB is non-relational database service designed for performance, scalability, and availability. Amazon DocumentDB decouple storage and compute that allow storage and compute to scale independently, and you can increase the read capacity to millions of records per second by adding up to 15 low latency read replicas in minutes, regardless of the size of your data. DocumentDB is designed for 99.99% of availability and replicates six copies of your data across AWS availability zones (AZs). Amazon DocumentDB allowing you to use your existing MongoDB drivers and tools. DocumentDB automatically and continuously monitors and backup your database to Amazon S3, by enabling point-in-time recovery up to the second for last 35 days. Amazon DocumentDB use a distributed, fault-tolerant, self-healing storage system that auto scale up to 64TB per database cluster. DocumentDB provides twice the throughput of currently available MongoDB managed services.

Amazon Neptune is fast, reliable, fully managed database that make it easy to build and run applications that works with highly connected datasets. Amazon Neptune is optimized for storing billions of relationships and querying the graph with millisecond latency. Amazon Neptune supports Gremlin and SPARQL open graph API and provides high performance for both of these graph models and their query languages. Neptune supports up to 15 low latency read replicas across three availability zones to scale and read capacity and execute more than one hundred thousand graph queries per second. Neptune is ACID compliance that provides more than 99.99% availability. Neptune replicates six copies of your data across three availability zones. Neptune continuously backs up your data to Amazon S3 and transparently recovers from physical storage failure. Neptune is serverless database, meaning, you do not need to worry about the server provisioning, patching, configuration, backup, and Monitoring. The use case of Neptune is Social networking, Recommendation Engine, fraud detection, etc.

Amazon Keyspaces is highly available, scalable, eventual consistence, and managed Apache Cassandra compatible database service. If you have existing Apache Cassandra workload on your data center, you can run the same Cassandra workload on AWS with the same application code and developer tools. Keyspaces is serverless, meaning you do not need to provision the server, no need to worry about software patch, and you do not need to maintain as well. Keyspaces provides performance as single digit millisecond response time and offers 99.99% availability SLA and default encryption of tables. You can build application using open source Cassandra API and drivers.

Amazon Timestream:

Amazon Timestream is timeseries database. You can put this in NoSQL database type but timeseries databases are slightly different and it is a new type of database provided by AWS. Timeseries databases are being used in IoT applications, DevOps, and Industrial Telemetry. Timestream is fast and fully managed database where you can easily store and analyze trillions of events per day at one tenth of the cost of relational database. Timestream is a time series database that can store and analyze events generated by wide range of IoT devices and smart industrial machines. In this database data comes in time order. Timestream database stores timeseries data that can only be appended. You can query Timestream timeseries database with the time interval as an input. Timestream is optimized for wide variety of time interval, like millisecond, microsecond, and nanoseconds.

Fig: How Timestream works.

Timestream is serverless database, meaning, you do not have to worry about server provisioning, software patching, scaling, and maintenance.

Amazon QLDB:

Amazon Quantum Ledger Database (QLDB) is fully managed ledger database that provides immutable and cryptographically verifiable transaction log owned by a central trusted authority. You can track each and every data changes. QLDB maintains complete history of changes over time. This database can be used for ledger or ledger like functionality, such as tracking of movement of an item in supply chain product or tracking of account activity in banking system. Benefit of using QLDB in such cases over relational database is relational database is not immutable and any update of data is hard to track, QLDB is immutable, so every time you update any data, a new data will get created. You have track of old data and sequence of update. Immutability is the biggest advantage of QLDB over relational database. QLDB uses an immutable transaction log known as journal, that track each application data change and maintain a complete and verifiable sequenced history of changes over time. You can easily access full history of database and analyze the history to see how your data has changed over time. QLDB uses a cryptographic hash function to generate a secure output file of change history of your data called digest. The digest act as a proof of your data change history. QLDB is highly scalable that can execute two to three times more transactions than ledgers in common blockchain frameworks. QLDB is serverless, meaning you do not need to worry about server provisioning, software patching, define capacity or read and write limits. You just need to create a ledger, define your tables, and QLDB will automatically scale based on the demand or load on your application. QLDB supports PartiQL, an open source SQL compatible query language. QLDB supports document-oriented data model that enable you to easily store and process structured and semi-structured data. QLDB transactions are ACID complains.

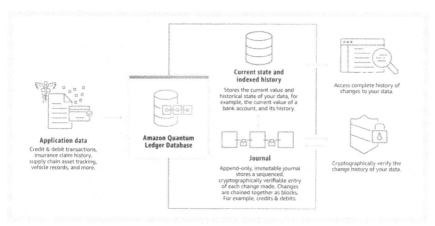

Fig: working of QLDB

You can stream your QLDB data directly to Amazon Kinesis Data Streams. Streaming your data to Kinesis allows you to react quickly to new events, you can integrate it with Lambda functions, you can easily develop event driven workflow to perform real time and historical data analysis.

In-memory Database:

In-memory database can be used for cache, session management, gaming leaderboards, etc. AWS has Amazon ElastiCache as In-memory database. ElastiCache comes in two version, ElastiCache for Memcached, and ElastiCache for Redis.

Amazon ElastiCache for Memcached is fully managed in-memory Memcached compatible key-value pair data store. This is frequently used for web and mobile app to store frequently accessed data in in-memory.

Fig: Working of ElastiCache for Memcached.

ElastiCache for Redis is fast, in-memory Redis compatible data store that fully supports open source Redis in-memory database. You can store frequently used data in this in-memory database. If you already have Redis database in your on-premises, you can easily setup the same in cloud.

Fig: Working of ElastiCache for Redis.

Databases in Azure:

Azure provides wide range of data store and data store API options for relational, NoSQL, In-memory, and Time Series databases.

Relational Database Service:

Azure database is relational database service available for MariaDB, MySQL, and PostgreSQL. Azure database service for any of these three databases provides features like, built-in availability, predictable performance, scale as needed within seconds, and automatic backup and point-in-time restore for up to 35 days.

Azure Database for MariaDB is fully managed relational database service for MariaDB. MariaDB is open source database. You can use Azure Database for MariaDB for fully managed PaaS that provides 99.99% availability. You can have Azure VM and deploy MariaDB on it that will provide you 99.95% availability with two or more instance within same availability set, or 99.99% availability using availability zones with multiple instances in multiple availability sets. If you already have MariaDB running in your on-premises server, you can easily migrate to the Azure Database for MariaDB.

Azure Database for MySQL is a fully managed relational database as a service for MySQL database. You can develop application with Azure Database for MySQL, and you will get advantage of using open source tools and platform. If you have existing MySQL database in on-premises sever, you can migrate easily to Azure Database for MySQL PaaS offering.

Azure Database for PostgreSQLL is a fully managed relational database as a service based on PostgreSQL database. This is available in two deployment options, single server and Hyperscale cluster. The Hyperscale cluster option horizontally scale queries across multiple machine using sharding and serves applications that requires greater scale and performance. You can easily move your on-premises PostgreSQL database to the Azure Database for PostgreSQL.

Azure SQL database:

Azure SQL database is fully managed PaaS database engine based on Microsoft SQL Server database. It is fully managed PaaS meaning, you do not need to worry about most of the database management functions, like patching, upgradations, backup, and monitoring. Azure SQL database is always running on latest stable version of Microsoft SQL server database and latest patch of operating system with 99.99% of availability.

You can define amount of resources defined for Azure SQL database with the option of single database and elastic pool. In single database each database is isolated from each other and is portable.

Each of these will have guaranteed amount of memory, compute, and storage resources. You can dynamically scale single database resources up and down. The Hyperscale service tier for single database will enable you to scale up to 100 TB with fast backup and restore capabilities. In case of Elastic Pool, you can get resources that are shared by all databases in the pool. You can move existing database to the resource pool to maximize the use of resources and save money. Elastic pool option gives you the ability to dynamically scale elastic pool resources up and down.

Azure SQL database has three purchasing model, the vCore-based purchasing model, the DTU-based purchasing model, and the serverless model. You can choose number of vCore, amount of memory, and amount and speed of storage in vCore-based purchasing model. DTU-based service model differentiated by range of compute sizes, with a fixed amount of included storage, fixed retention period for backups, and fixed price. The serverless model automatically scales compute based on the workload demand.

Azure SQL database offers three service tiers that can be chosen based on the type of application. General purpose or standard service tier provides budget oriented balanced compute and storage options for general purpose workload. Business critical or premium service tier is designed for OLTP applications with high transaction rate and low latency I/O. Hyperscale service tier is designed for very large OLTP database and ability to auto scale storage and compute gracefully.

Azure Cosmos DB:

Azure Cosmos DB is a NoSQL database that is globally distributed, multi-model database service. Multi-model database meaning you can access your data using SQL, MongoDB, Cassandra, Tables, or Gremlin database API. Cosmos DB provides high availability, high throughput, low latency, and consistency. Cosmos DB provides you five consistency options to choose from based on the need of your application, those five consistency options are, Strong, Bounded staleness, Session, Consistence prefix, and Eventual consistency. Cosmos DB guarantees less than 10 millisecond latencies both read (indexed) and write at the 99th percentile around the world. No schema or index management is required therefore you do not have to worry about application downtime while migrating schemas. You can use Cosmos DB for any web, mobile, gaming, and IoT applications. If your application has to handle massive amount of data, reads, and writes at global scale with near real time response time for a variety of data, Cosmos DB will be a good choice to consider. Cosmos DB provides the choices of APIs you can use with, SQL is chosen by default if you don't have any choice. Below figure shows the Cosmos DB database types and type of API can be used for data access.

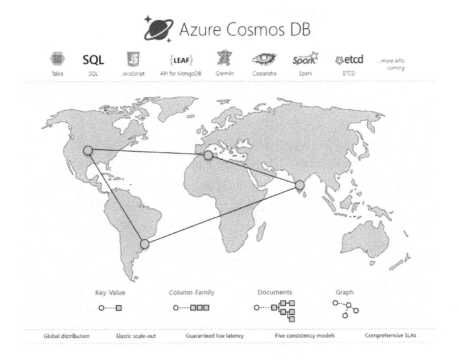

Azure Table Storage:

Azure table storage is a service that stores structured NoSQL data in cloud. Azure table storage is key/attribute store with a schemaless design.

You can use table storage if you need to store terabyte of structured schemaless data that you can access quickly. You can store user data for web applications, address book, device information, or other type of metadata your application need to store and fetch.

Table in Azure table storage is a collection of entities, that does not enforce schema on entities. That means a single table can have entities having different set of properties. Entity is a set of properties similar to a database row. Properties is a name value-pair, each entity can have up to 252 properties.

Databases in Google Cloud:

Google cloud provides wide range of relational, NoSQL, timeseries, in-memory and analytical data store. It provides flexible performance and scale with broad open-source capability. Here are databases provided by Google Cloud.

Cloud SQL:

Cloud SQL is fully managed relational database service for MySQL, PostgreSQL, and SQL Server. Fully managed meaning, you do not have to maintain and patch database server or operating system, All security patch, software patches, software upgrades, and database upgrade will happen automatically. It also takes care about backups, replication, encryption patches, and capacity increases. It fully integrated and can connect easily with other google cloud services.

Cloud SQL is ideal solution for website and mobile backend, game states, CRM tools, MySQL, PostgreSQL, and Microsoft SQL Server databases.

Cloud SQL for MySQL provides fully managed MySQL community edition database in the cloud. It currently supports 5.6 and 5.7, provides up to 416 GB of RAM and 30 TB of data storage with the option of automatically increase of storage size as needed. Customer data will be encrypted on Google's internal network, database tables, temporary files, and backups. It supports for secure external connection with cloud SQL proxy, or with SSL/TLS protocols. Your data will be replicated in multiple zones and provides automatic failovers. It provides supports for MySQL wire protocol and standard MySQL connectors.

Cloud SQL for PostgreSQL is fully managed PostgreSQL databases in the cloud. You can select custom machine type with 416 GB of RAM and 64 CPUs. It is based on the Cloud SQL second generation platform. You can have up to 30 TB of storage with the ability to automatically increase storage size as needed. Your data will be encrypted on Google's internal networks and in the database tables, temporary files, and backups. It has support for secure external connections with the cloud SQL proxy or with the SSL/TLS protocol. Point-in-time recovery and import/export in csv format is not available in Cloud SQL for PostgreSQL.

Cloud SQL for SQL Server is fully managed SQL Server database in the cloud. It supports custom machine type with up to 416 GB of RAM and 64 CPUs. You can have up to 30 TB of storage with ability to automatically increase of storage size as needed. Your data will be encrypted on Google's internal networks, database table, temporary files, and backups. It supports for secure external connections with the cloud SQL proxy or with the SSL/TLS protocol. SQL Server reporting services (SSRS), SQL Server analysis services (SSAS), and SQL Server integration services (SSIS) will not be available through Cloud SQL, but you have separate compute engine instance for these services. Also AD authentication and WCF data services will not be available through Cloud SQL.

Cloud Spanner:

Cloud Spanner is fully managed, scalable, relational database service for regional and global application data. Cloud Spanner is unique relational database that provides strong consistency, horizontal scaling, and high availability with enterprise grade security.

Cloud spanner provides structural database structure with schema, SQL support, and strong consistency similar to the traditional relational database, it provides high availability, and horizontal scalability similar to the NoSQL database. That means Cloud Spanner is the database provides you features of Relational and Non-Relational database in a relational database service. You do not have to configure the replication, replication in Cloud Spanner is automatic.

Cloud Spanner provides strong consistency for all transactions. Spanner is also a highly available global scale distributed database. As per CAP theorem the combination of availability and consistency over the wide area is generally considered impossible. But Cloud Spanner achieve this combination.

Cloud Firestore:

Cloud Firestore is fully managed, serverless, cloud native, document-based NoSQL database. Firestore simplifies storing, syncing, and querying data from mobile, web, or IoT application at global scale. Firestore is integrated with Google Cloud Platform and Google's mobile development platform Firebase.

Cloud Firestore provides live synchronization, offline support, and ACID transactions. Offline support meaning, you can make changes of your data while you are offline, and your changes will be synced to the cloud when you come back online. For example, your mobile application is talking with the Firestore and if you did any modification while there is no network and you are offline, your changes will get synced once you will get network and you becomes online.

Cloud Bigtable:

Cloud Bigtable is fully managed, highly scalable NoSQL database service for large analytical and operational workload. It provides consistent 10 millisecond latency can handle millions of requests per second. You will be able to reconfigure Bigtable without having any downtime.

Cloud Bigtable is ideal for storing very large amount of single-keyed data with very low latency. It provides very high read and write throughput at low latency. Bigtable store data in key-value pair where each value should less than 10 MB.

Bigtable is good for the application that needs very high throughput and scalability, MapReduce operations, stream processing, analytics, and machine learning. You can use Bigtable to store and query time-series data, marketing data like purchase history, financial data like transaction history, IoT data like uses reports, and graph data like how users are connected to each other.

Cloud Memorystore:

Cloud Memorystore is in-memory database service for Redis and Memcached. Memorystore is highly scalable, secured, and highly available. Memorystore use to build application cache that provides sub-millisecond data access and is fully compatible with open source Redis and Memcached. You can migrate your existing caching layer to the Memorystore with zero code change.

Memorystore is fully managed meaning you do not have to worry about resource provisioning, replication, failover, and software upgradation and patching. Memorystore provides 99.99% availability SLA with automatic failover that ensure that your application will be highly available. Memcached supports multithreaded architecture and designed for simplicity, otherwise you will get everything in Redis that is provided by Memcached.

BigQuery:

BigQuery is serverless, highly scalable data warehouse that can analyze petabyte of data using ANSI SQL at very high speed and at low cost.

BigQuery ML provides the capability to create and execute machine learning models in BigQuery using standard SQL queries.

You do not have to program machine learning solution using Python or Java, Models are trained and accessed in BigQuery using SQL. BigQuery ML query execution time is not dependent on the volume of data you have, it provides constant time of query.

BigQuery BI Engine is very fast in-memory analysis service for BigQuery that allow user to analyze large and complex dataset interactively with sub-second query response time and high concurrency. It can easily integrate with the DataStudio to accelerate data exploration and analysis.

BigQuery GIS is the serverless architecture of BigQuery with native support for geospatial analysis. You may record latitude and longitude of your delivery vehicle or package over time. BigQuery GIS lets you analyze and visualize geospatial data in BigQuery by using geography data types and standard SQL geography functions.

You can compare BigQuery with Amazon Redshift data warehouse or Teradata. Using Google BigQuery Data Transfer service, you can transfer your data from Teradata data warehouse or Amazon Redshift data warehouse.

Index

B

C

D

E

F

G

R

S

T

V

W

www.ingramcontent.com/pod-product-compliance
Lightning Source LLC
Chambersburg PA
CBHW071253050326
40690CB00011B/2372